D0216407

PARTICIPATORY GOVERNANCE

Participatory Governance

Planning, Conflict Mediation and Public Decision-Making in Civil Society

Edited by

W. ROBERT LOVAN
MICHAEL MURRAY
RON SHAFFER

ASHGATE

Published by
Ashgate Publishing Limited
Gower House
Croft Road
Aldershot
Hants GU11 3HR
England

Ashgate Publishing Company
Suite 420
101 Cherry Street
Burlington, VT 05401-4405
USA

Ashgate website: http://www.ashgate.com

British Library Cataloguing in Publication Data
Participatory governance : planning, conflict mediation and
public decision-making in civil society
1.Political planning - Citizen participation 2.Political
participation 3.Social contract 4.Comparative government
I.Lovan, W. Robert II.Murray, Michael R. III.Shaffer, Ron
323'.042

Library of Congress Cataloging-in-Publication Data
Participatory governance : planning, conflict mediation and public decision-making in civil society / edited by W. Robert Lovan, Michael Murray, and Ron Shaffer.
p. cm.
Includes bibliographical references and index.
ISBN 0-7546-1852-8
1.Public administration--Citizen participation--Cross-cultural studies. I. Lovan, W. Robert. II. Murray, Michael, 1953- III. Shaffer, Ron, 1945-

JF1351.P28 2003
323'.042--dc21

2003052488

ISBN 0 7546 1852 8

Reprinted 2005

Printed & bound by MPG Books Ltd, Bodmin, Cornwall

Contents

SECTION THREE: OPERATIONAL DIMENSIONS OF PARTICIPATORY GOVERNANCE

SECTION FOUR: REFLECTIONS ON PARTICIPATORY GOVERNANCE

List of Figures

List of Tables

Acknowledgments

The completion of this book has been some four years in the making. It has its origins in a seminar convened in Baltimore, USA in September 1998 at which an invited group of policy advisers, practitioners and academics with a shared interest in rural development met for a two day discussion on the theme *New Governance at the Millennium: Transforming Public Sector Decision-Making*. We would like to express our gratitude to our colleagues who contributed so many insights during that seminar on the broad dynamics and operational principles shaping new governance. In particular, we wish to formally acknowledge the financial support received for this seminar from the University of Wisconsin / Madison-Extension, the Farm Foundation (Chicago), the four Regional Rural Development Centers at Oregon State University, Iowa State University, Mississippi State University and Pennsylvania State University, and the Rural Policy Research Institute at University of Missouri.

When the guiding theme for this book was subsequently agreed we were pleased to secure a publishing contract with Ashgate. We would like to thank Valerie Rose and Carolyn Court at Ashgate for their enthusiastic support and continuous patience in the run-up to publication.

There would be no book, of course, without contributors and in this regard we as editors are appreciative of the willingness showed by everyone we approached to respond by way of a chapter. The richness of the narratives reflects a deep commitment by our colleagues to participatory governance and we are grateful that each contributor has engaged admirably with the shared challenge of critical reflection on our collective practice. We are grateful to Killian McDaid at University of Ulster for preparation of the chapter illustrations.

Finally, we wish to express our thanks to the Rural Innovation and Research Partnership (RIRP) in Northern Ireland for a grant award towards the publication costs of this book.

W Robert Lovan
Michael Murray
Ron Shaffer

List of Contributors

Jim Cavaye is a practitioner, educator and researcher in community and regional development with 20 years experience working with rural communities and government agencies. He is Director of Cavaye Community Development based in Queensland, Australia. He has a PhD in community development from the University of Wisconsin and is an Adjunct Associate Professor at the Institute for Sustainable Regional Development in Central Queensland University and at the Community Service and Research Center, University of Queensland.

David Deshler is Professor Emeritus at Cornell University. His work has focused on adult and community education, community development processes, participatory action research methods, and international development education approaches at the Department of Education, New York State College of Agriculture and Life Sciences. He was faculty leader for the Natural Resource Management and Sustainable Agriculture Program in Ghana (1994-2000), a partnership effort of the Cornell International Institute for Food, Agriculture and Development. He received his Ed.D degree from the University of California, Los Angeles.

Frank Dukes is Director of the Institute for Environmental Negotiation (IEN), University of Virginia where he is responsible for the design of dispute resolution and public participation processes, mediation and facilitation, teaching and training, and conducting research. His book *Resolving Public Conflict: Transforming Community and Governance* describes how public conflict resolution procedures can assist in vitalizing democracy. He is also co-author of *Reaching for Higher Ground in Conflict Resolution* which describes how diverse communities and groups can create expectations for addressing conflict with integrity, vision and creativity. He initiated the Community-based Collaborative Research Consortium (cbcrc.org) and is lead author of *Collaboration: A Guide for Environmental Advocates* with the Wilderness Society and the National Audubon Society. For seven years he mediated discussions involving tobacco farm and public health leaders, resulting in a series of agreements including over $2 billion in funding for tobacco farm regions and tobacco control efforts,

Kirby Edmonds lives in Ithaca, New York and is the managing partner of Training for Change Associates, a human relations training and consulting firm based in Ithaca. He is a management consultant, mediator and trainer and has over 30 years experience helping groups to build and manage networks, deal with conflict, create consensus and solve problems. He has been a consultant to public and private partnerships throughout the United States. He received his BA and Masters degrees from Cornell University.

Cornelia Butler Flora is the Charles F. Curtis Distinguished Professor of Agriculture and Sociology at Iowa State University and Director of the North Central Regional Center for Rural Development in the United States. She is a past president of the Rural Sociological Society, president of the Agriculture, Food and Human Values Society, president of the Community Development Society and a fellow of the American Association for the Advancement of Science. She is currently on the Board of Directors of the Heartland Institute for Community Leadership, the Board on Agriculture of the National Research Council of the National Academy of Science, the Midwest Assistance Program, the Northwest Area Foundation and Winrock International. She is author and editor of a number of recent books and over 185 articles and book chapters primarily on rural development in the United States and developing countries. She and her husband, Jan L. Flora, continue research in the Andean region of Latin America.

Richard Gardner is the principal consultant at Bootstrap Solutions, a firm specializing in rural development, economics, strategic planning and group facilitation. From 1992 through to July 2001 Dr Gardner was the Executive Director of the Idaho Rural Partnership prior to which he was a policy economist for nine years in the Idaho Division of Financial Management working on agricultural, natural resource, rural development and tax issues. He has published widely, has participated in over 200 workshops on rural development and has facilitated the creation and delivery of numerous rural programs and projects. In 2000 he received the USDA's top achievement, the Honor Award for Public Service, from Secretary Dan Glickman and in 2001 he was presented with the Ron Shaffer Award for Collaborative Leadership by the National Rural Development Partnership.

John Greer is Senior Lecturer in the School of Environmental Planning, Queen's University, Belfast, Northern Ireland. He is a past President of the Irish Planning Institute. His teaching and research interests include rural planning and development, planning theory and history of settlement. He has published numerous articles and research reports and is co-editor of *Rural Development in Ireland* (1993, Avebury) and *Rural Planning and Development in Northern Ireland* (2003, Institute of Public Administration, Dublin). He has recently completed a major research project on rural services provision as part of a wider commitment by Government to rural proofing.

Christian Huillet (1940-2002) retired from the Organization for Economic Cooperation and Development (OECD), Paris, France as the Principal Administrator – Technical Cooperation Committee. Christian, a French citizen, had a distinguished career as an international civil servant. He worked with developing countries in North Africa, with the OECD inter-governmental Early Development Programs in Southern Europe and the Mediterranean countries, and directed OECD policy studies in Australia, Japan, North America and both western and eastern Europe.

Michael Kenny is Lecturer in community and rural development at the Center for Adult and Community Education at NUI Maynooth in Ireland and is Coordinator of an inter-university MSc program in rural development. He has extensive experience in assisting communities in local development participation. His most recent publication is a report for the Department of Agriculture, Food and Rural Development titled *Rural Regeneration through Universal and Equitable Public Service Delivery*.

Carol Kuhre has served as Executive Director of Rural Action in the United States since 1990. During that time she has worked with members to develop and execute a strategy for rural renewal for the Appalachian counties of Ohio. Prior to serving with Rural Action she worked as a campus minister with Lutheran Campus Ministry at Penn State University and Ohio University. For 15 years she was the Co-Director of United Campus Ministry, an ecumenical ministry at Ohio University in Athens, Ohio. Carol holds degrees from Concordia College and Ohio University and has also studied at the Lutheran School of Theology in Chicago. Carol, with her husband, Dr Bruce Kuhre, has traveled and lectured on the interface between art, theology and social change in Botswana, the Philippines, Belize and Austria. Her avocation is fiber arts and she owns Cabin Crafts Handweaving and is a founding member of Athens Tapestry Works.

W. Robert Lovan is a senior level federal program manager with the United States Department of Agriculture in Washington, DC and is responsible for providing mediation services for Rural Development. He has been responsible for developing and delivering a number of community development programs throughout the United States and has directed work on two Presidential initiatives on the effective utilization of public and private resources at the local level. The most recent initiative involves the implementation of public-private rural partnerships, which are based on individual negotiated agreements between State Governors and the national government. To date three books have been published dealing with this intergovernmental project. His federal career includes work at local, state and national levels, a University of Arizona faculty member, and international rural development work in Southeast Asia and with the OECD in Western Europe. He is currently completing a doctorate at George Mason University on conflict analysis and resolution with an emphasis on education and training in the public and private sectors.

Jeanne Meldon is a planning and environment consultant in Ireland and is a member of the Irish Planning Institute. She has worked on a wide range of projects, a number of which have involved trans-national partners, dealing with local and regional development, EU Structural Funds, spatial planning, sustainable development, tourism and the environment, and landscape management. During the earlier part of her career she worked as an Assistant Planner with Galway, Wicklow and Kildare County Councils. She has been a member of the National Economic and Social

Forum of Ireland since 1993 and a member of COMHAR (The National Partnership for Sustainable Development) since its inception in 1999.

Michael Murray is a Reader in the School of Environmental Planning at Queen's University, Belfast, Northern Ireland from where he received his PhD. His research interests include partnership governance, strategic planning and community-led rural development on all of which he has published widely. His research activity takes him regularly to the United States where he has been a Visiting Scholar at Colorado State University engaged in analyses of federal-state partnership activities and community regeneration. He is the author of *The Politics and Pragmatism of Urban Containment* (1991, Avebury), co-author of *Revitalizing Rural America – A Perspective on Collaboration and Community* (1996, John Wiley & Sons) and *Partnership Governance in Northern Ireland* (1998, Oak Tree Press), and co-editor of *Rural Development in Ireland* (1993, Avebury) and *Rural Planning and Development in Northern Ireland* (2003, Institute of Public Administration, Dublin).

Etienne Nel is Associate Professor in Geography at Rhodes University, Grahamstown, South Africa where he also received his PhD. His research focuses on urban and local economic development in South Africa and southern Africa and he has published two books including *Regional and Local Economic Development in South Africa* (1999, Ashgate) as well as several dozen articles and book chapters in this field.

Randall Prior is an Episcopal priest and Rector of St Andrew's Episcopal Church, Burke, Virginia in the United States. His background includes extensive work as a pastoral counselor and teacher of pastoral theology. He has been instrumental in developing educational and community service programs during the course of his ministry, which have extended to his current service as President of Habitat for Humanity of Northern Virginia. He has also assisted efforts in developing overseas partnerships between his own congregation and those in Africa and the Middle East. He is an adjunct faculty member in the Pastoral Theology Department of the Virginia Theological Seminary, Alexandria, Virginia.

Norman Reid is Deputy Administrator, Office of Community Development, Rural Development, United States Department of Agriculture, which manages community development delivery systems, provides technical assistance and information to USDA's field staff, and implements the rural Empowerment Zones / Enterprise Communities program. For a number of years he conducted research on rural development issues and is author of over 90 publications. He has staffed three national rural development commissions and served as consultant to the OECD in Paris. In 1975 he was awarded a PhD in Political Science from the University of Illinois-Urbana. In 2000 he received the Kenneth Wilkinson Rural Policy Award from the Rural Policy Interest Group of the Rural Sociological Society.

Christopher S. Rice is Research Director at the University of Kentucky Appalachian Center. He received his PhD in Political Science from the University of Kentucky in 2002. His dissertation *Discourses of Sustainability: Grassroots Organizations and Sustainable Community Development in Central Appalachia* is an ethnographic analysis of the formation of discourses of sustainability by NGOs in three Appalachian counties in Kentucky, Ohio and Virginia and the programs and policies of sustainable development that emerge from this process. It also generates a set of general principles for NGO success in struggles for community sustainability. Christopher's writing attempts to blend the development of critical political theory with the construction of political ethnographies concerning issues of space and place. His next project will be a manuscript examining the American comic book as critical political space.

Ron Shaffer is Emeritus Professor of Agricultural and Applied Economics at the University of Wisconsin-Madison in the United States where he has taught and researched in the area of Community Economics. This included teaching a graduate level course on Community Economic Analysis and helping to organize a new Masters program on Community Development. He was a faculty member at the University of Wisconsin-Madison from January 1972 through to October 2001, served as Director of the University of Wisconsin Center for Community Economic Development from July 1990 through to June 2000, and directed the National Rural Economic Development Institute, which was part of the National Rural Development Partnership, from 1990 to 1998. He has worked with the Rural Policy Research Institute on national rural development policy analysis since 1993. His extension efforts have emphasized working with communities in Wisconsin to create economic development strategies.

Robert Stoker is Associate Professor of Political Science at George Washington University. The George Washington Institute for Public Policy recently selected him as a Policy Research Scholar. His research interests include US social policy, urban redevelopment, and policy implementation. He is the author of *Reluctant Partners: Implementing Federal Policy* (1991, University of Pittsburgh Press) and has published numerous articles on United States politics and public policy issues.

Jim Walsh is Professor and Head of the Department of Geography at NUI Maynooth in Ireland. He is a founding member of the National Institute for Regional and Spatial Analysis. He has published extensively on a wide variety of topics related to regional and rural development and spatial planning. Recent books include *Regional Planning and Development in Europe*, 2000, Ashgate (co-edited with D. Shaw and P. Roberts), and *Irish Agriculture in Transition – A Census Atlas*, 1999 (with S. Lafferty and P. Commins). He has undertaken many evaluations of local development programs and has become involved in preparing strategies for the integration of local government and local development.

Foreword

The term *governance* goes beyond the discussion of public management to a more fundamental question of how the processes of democracy (citizen involvement, decision-making procedures and administrative functions) can be adapted to help countries resolve the complex public issues with which they are challenged. In the Organization for Economic Cooperation and Development (OECD) governance is defined in terms of relationships. The instruments and methods of governance include more than just the functions of public administrators in managing the structures of public institutions. Governance encompasses the many and ever changing sets of relationships between the government and the interests of citizens, who interact with public institutions both as individuals and as participants with mutual interests.

Governance is frequently employed in political and academic dialogue, but its usage does lack a precise definition. In the French language, for instance, there is no single word that carries the same multiple connotations as in the English language. For many countries, governance is defined as the sum of the many possible organisational mechanisms and the types of relationships by which individuals interact with their public sector institutions in meeting the needs of society. The diversity implicit in these configurations is exemplified as follows:

- In the *Napoleonic* countries (France, Italy, Spain, Belgium) and to a certain extent the southern European countries (Portugal, Greece, Turkey) the exercise of governmental responsibility is based on legal traditions. Because there is a tradition of distrust between the levels of government, it is believed necessary in these countries to separate the types of services and the accompanying tasks and assign each service and the task to a different level of government. This separation is also seen as a type of guarantee of efficiency. Priority is given to the coherence of the institutional framework. It is believed that good governance should flow from that coherence. The citizen is traditionally looked upon as a subject, although this view is changing. Incorporating different possible mechanisms in government is gaining ground for the delivery of services that have traditionally been provided by public enterprises, not by private institutions.
- The *Germanic* countries (Germany, Austria, Switzerland) are also heirs to a legal culture grounded in a respect of the law. This respect determines much of the context for governance. There is, however, a heavy reliance on co-operative, intergovernmental mechanisms to ensure that the expected public responsibilities are carried out by each layer of government. Local authorities are entitled by statute to use a variety of administrative structures in providing the services for which they are responsible, but the mechanisms

can be flexible in adjusting to changing circumstances. The local level, however, is largely regulated by the central government. Those services that exceed the capacity of the local municipalities are entrusted to intermediate levels. The intermediate levels of government do play an important role in the federal legislative process. Switzerland is, however, a special case. Local democracy is very strong with the cantons holding and using a large amount of power, both financially and administratively. Citizens are given a considerable degree of responsibility to oversee the administrative mechanisms and the operating institutions to deliver the expected public services. Each of the cantons determines its own level of public service provision, as well as the appropriate administrative structure and the necessary procedures for delivering the expected public services.

- The *Anglo-Saxon* countries (United States, Canada, Australia, United Kingdom, Ireland) place a particular emphasis on efficiency, effectiveness and value-for-money. They are more likely to introduce alternative administrative mechanisms, for example, partnerships, in order to carry out public responsibilities. The citizen is viewed primarily as a customer or client for public services.
- The *Nordic* countries (Denmark, Finland, Norway, Sweden) belong to a political culture concerned with meeting the needs of citizens. Their tradition is one of negotiation and consultation between the institutions of government and the citizens. Central government regulation has, on the whole, been eased with the aim of giving local levels the possibility of doing a better job of adjusting to local conditions, as well as using their finances in accordance with local priorities. An important part of the approach is the ability of local authorities to organize their funds according to local needs.

Governance, therefore, is about processes of making decisions. In other words, it is concerned with processes focussing on the distribution of public responsibility across multiple stakeholders. In dealing with this theme in contemporary societies which span Africa, Australia, Europe, and the United States, this book, titled *Participatory Governance: Planning, Conflict Mediation and Public Decision-Making in Civil Society*, offers important tools. First, is a model for gaining insights into the dynamics of participatory governance. Second, is a type of road-map for identifying and assessing the component parts of participatory governance. Third, is an insight into alternative relationships between citizens and their public institutions. These very informative chapters, written by an international grouping of practitioners and academics, deserve serious reflection on what can work differently and better.

Christian Huillet
Masson Laffitte, France

Section One:
Introduction

Chapter 1

Participatory Governance in a Changing World

W. Robert Lovan, Michael Murray and Ron Shaffer

Introduction

The past fifteen years have evidenced a growing and diverse literature on how the work of government can be better secured. Public administration has embraced a remedial vocabulary of *New Public Management*, *Reinventing Government*, and *New Governance*. In a world where big business has become a global phenomenon, the need to reform bureaucracies and develop performance based public sector organizations has similarly crossed international borders, albeit perhaps that the receptiveness of innovation in this sphere is greatest in the United States, the United Kingdom, Canada, Australia and New Zealand. This stirring of consciousness has been no better illustrated than by the analysis of Osborne and Gaebler in their 1992 book *Reinventing Government: how the entrepreneurial spirit is transforming the public sector*. At the time, this established itself as a bestseller with its messianic tone capturing the *Zeitgeist* of politicians and practitioners that government should seek to become more catalytic, community owned, competitive, mission driven, results oriented, customer driven, enterprising, anticipating, decentralized, and market orientated. Notwithstanding the fundamental criticisms that there are limitations to a business-based paradigm derived from public accountability and public service imperatives, the debates on governance restructuring, which this book in part facilitated, have generated a momentum for change. There is now, as argued by Roberts (1999), a shared appreciation of the need to prune non essential programs, give managers more freedom to use resources wisely, focus on results rather than inputs, and rely more on the private sector for service delivery.

However, the gap between the desire for reform and actual achievement varies from country to country and, arguably, the scope for action can be interpreted as a reflection of prevailing institutional and political complexities. Thus, for example, the difficulties faced in securing implementation of the recommendations of the National Performance Review brought forward in 1993 by the Vice President of the United States, Al Gore, were inextricably linked to the need for legislative approval. Moreover, this initiative by seeking to challenge at that time the traditionally strong management role of Congress, melt the boundaries between public organizations and give federal workers greater decision-making authority was also obliged to confront the resistance driven by internal organizational political dynamics related to power, control and turf relationships (Thompson and Ingraham, 1996). Much the same analysis can be set against the more recent difficulties associated with the passage of legislative arrangements for faith-based initiatives in the current Bush administration. It is within this context of constrained change that the observations of Kettl (1999) are especially apt:

> The US has adopted some of the most aggressive and ambitious and sweeping efforts compared to all the other nations. It has launched more battles on more fronts than virtually any other country. On the other hand, our system of government makes it difficult to initiate and sustain that kind of change. By almost anybody's measure, we have a more complicated environment to work in.

The important point to make is that there is no single policy and operational model at work and that preferences for various possibilities are teased out through the interplay of market orientation and managerialism. As illustrations of the former, Hansen (2001) suggests the adoption of privatization, contracting out, purchaser-provider models, free choice/exit opportunities (for example, pension schemes) and competitive based incentives; these have the effect of pushing public administration more towards self-regulating and non-political mechanisms of social co-ordination. On the other hand, efforts to transform the public sector on the basis of leadership and management practices developed in the private sector point to the introduction of service and quality performance systems (for example, Total Quality Management, benchmarking), the decentralization of decision-making competence and responsibility, and greater emphasis on monitoring and evaluation. In other words, the traditional conceptualization of the public sector, whereby national governments are the major actors in public policy is in doubt (Peters and Pierre, 1998). A large number of other stakeholders are active in policy and administration arenas, and while some would argue that, accordingly, the traditional conceptualization of central government as a controlling and regulating organization for society is outmoded (Bekke *et al*, 1995), it may be more appropriate to hold that the restructuring observed above is rather more evolutionary than revolutionary. The shift is towards greater interdependency, for example, between central governments, or between central and regional (or state) government and other sub-national actors.

As important as mapping the organizational configurations is, mapping alone provides insufficient understanding of the underlying dynamics associated with this adjustment. Accordingly, this introductory chapter commences by exploring key elements driving the processes of change within the arena of public decision-making. It is argued that the necessity to work through networks, give greater recognition to diversity and division in contemporary society, promote deeper responsiveness to service constituencies and engage in the reshaping of accountability relationships provide for an interpretative progression beyond conventional government and towards governance as a new way of governing. The leitmotif of this book is participation, and thus, against that backcloth, the chapter then moves on to review the concept of civil society as an organizing idea for participatory governance. The important point being made is that civil society is subject to multiple meanings. This complexity is subsequently distilled into an interactive model of participatory governance which seeks to capture the multi-dimensionality of public policy decision-making in civil society. The model is meant to be descriptive, offer a guiding framework for analysis and not be necessarily predictive. A set of key issues, drawn from the wider literature, are then rehearsed simply to demonstrate that participatory governance is not without its conceptual difficulties and operational constraints. The chapter concludes by briefly explaining the structure of the remainder of the book and, in particular, the role played by the series of in-depth case studies in identifying a set of robust principles for policy and practice.

From government to governance

Contemporary conceptualizations of governance place emphasis on interdependence between governmental and non governmental bodies, with the role of central government being reduced over time to one of seeking to co-ordinate or manage policy networks through facilitation and negotiation (Cloke *et al.*, 2000). The policy world is now made up of these diverse, overlapping and integrated networks which comprise varying relational geometries of the state, market and associational sectors.[1] The actors from each sector bring their own specific sets of power positions, roles and responsibilities as determined by values, skills and organizational resources into the network arena. Rhodes (1997) argues that these networks can vary along a continuum according to the closeness of the relationships within them. Thus at one end he identifies 'policy communities' characterized by (1) a limited number of participants with some groups consciously excluded; (2) frequent and high quality interaction between all members of the community on all matters related to the policy issues; (3) consistency in values, membership and policy outcomes which persist over time; (4) consensus, with the ideology, values and broad policy preferences shared by all participants; (5) resources to hand by all participants, including the delivery of compliant members by leaders; and (6) dominance perhaps by one participant, but this offset by all participants seeing themselves in a positive-sum game. At the opposite end of the continuum Rhodes identifies the 'issue network' which involves only policy consultation with the following characteristics: (1) many participants; (2) fluctuating interaction and access for the various members; (3) limited consensus and ever present conflict; (4) interaction based on consultation rather than negotiation and bargaining; and (5) an unequal power relationship in which many participants may have few resources, little access and no alternative.

The value of this network distinction for purposes of this book lies with its applicability to empirical investigation. It provokes critical questions over who is involved and not involved in governance arrangements, why this should be so, and what it is that they actually do. If government is perceived as 'power over', governance may be more correctly interpreted as 'power to', with stakeholders attempting to gain a capacity to act by blending their resources, skills and purposes into a viable partnership (Goodwin, 1998). Such ideas bring to mind Giddens (1986) theory of structuration which places emphasis on the relational webs within which we live as an explanation of societal change. Healey (1997) usefully summarizes this analysis as follows:

> We live through culturally-bound structures of rules and resource flows, yet human agency, in our continually inventive ways, remakes them in each instance, and in remaking the systems, the structuring forces, we also change ourselves and our cultures. Structures are 'shaped' by agency, just as they in turn 'shape' agency. Individual autonomy is thus always constituted in this interactive way, as, through the relations with others which we have, we individually and collectively maintain and change our social worlds. Structuring forces shape the systems of rules, the flows of resources and the systems of meaning through which we live. (p47)

At a more prosaic level this interconnectedness is recognition of what Bryson and Crosby (1992) have dubbed a shared power or a 'no-one in charge' world in which:

> no-one organization or institution is in a position to find and implement solutions to the problems that confront us in society. In this world, organizations and institutions that share objectives must also partly share resources and authority to achieve goals. (p4)

A second dynamic with a strong bearing on the shift from government to governance is greater recognition of the need to embrace diversity and the challenges of division in contemporary society. At a global scale various forms of difference continue to manifest themselves as antagonism and opposition whether these be in the form of nationality, ethnicity, language, religion or culture. At their most extreme the outcomes can be civil war or 'ethnic cleansing'. While relationships between territory and groups are one manifestation of diversity which can spill over into division, the group associations of individuals can also be a sphere of contestation. This may express itself as social prejudice which MacGreil (1996) defines as:

> a hostile, rigid and negative attitude towards a person, group, collectivity or category, because of the negative qualities ascribed to the group, collectivity or category based on faulty and stereotypical information and inflexible generalizations. (p19)

It is against these considerations that MacGreil identifies the manifestations of social prejudice as (1) ethnocentrism - conviction that members of other cultures or nationalities are inferior to one's own culture or nationality because of their culture, nationality or way of life; (2) political prejudice - rejection of people deemed to belong to certain political groups or categories, because of their political views; (3) religious prejudice based on perceived religious affiliation or non affiliation and encompassing sectarianism; (4) sexism which incorporates prejudice based on gender; (5) homophobia which focuses on attitudes and behavior related to sexual orientation; and (6) other social prejudice categories extending from ex-prisoners, people with mental and physical disability, people who are unemployed, to unmarried mothers.

The value of this classification, regardless of how the content may be debated, lies with its acknowledgment that we live in a society of complex group interactions. These interactions, MacGreil suggests, generate multiple responses and differentiated public policies. These can change relationships in a negative fashion, for example, by avoidance and withdrawal, fear and anxiety, expulsion and segregation, aggression and conflict. On the other hand interactions can be positive, say through a commitment to pluralism, whereby group differences are to be facilitated and respected, and there is equality of treatment for all groups.

In taking this forward, Healey (1996) has argued that there needs to be in place a variety of institutional mechanisms to develop some degree of social cohesion, as well as address some of the problems of social polarization and inequality. The key, she suggests, to effective institutional design includes fostering an interactive and collaborative capacity, building the social capital of trust and the intellectual capital of understanding based on discussions which allow the different views of diverse stakeholders to be opened up and explored. The implications for governance are cogently set out by Gould (1996) in her assertion that taking differences seriously in public life requires a radical increase in opportunities for participation not only in the discourse of the public sphere, but in the institutions of economic, social and political life, including its voluntary associations, social movements and informal groupings. Diversity, by inviting respect, thus becomes a positive social good.

Another key dynamic with a bearing on the governance process is the effort now made to enhance public policy responsiveness to service constituencies through citizen involvement mechanisms. Again the universality of this application is evidenced by the publication in 2001 of the OECD handbook *Citizens as Partners* which seeks to give best practice advice on information, consultation and public participation in policy making (Gramberger, 2001). In a similar vein the UK Government has published a raft of advice materials on the need for consultation and has also given advice on methods to be adopted. Thus, for example, in 1998 the Cabinet Office issued *An introductory guide: how to consult your users*. This was followed in 1999 by *Involving users: improving the delivery of local public services* and then in November 2000 by a *Code of Practice on Written Consultation*. The alleged benefits of consultation set out in this documentation are listed in Table 1.1 below.

Table 1.1 Benefits of Consultation

- Helps you plan services better to give users what they want, and expect.
- Helps you prioritize your services and make better use of limited resources.
- Helps you set performance standards relevant to users' needs and monitor them.
- Fosters a working partnership between your users and you, so they understand the problems facing you, and how they can help.

- Alerts you to problems quickly so you have a chance to put things right before they escalate.
- Symbolizes your commitment to be open and accountable: to put service first.

Source: UK Government Cabinet Office (1998) *An introductory guide: how to consult your users*, Chapter 1: How to consult your users.

The involvement techniques which have gained currency, as observed by Leach and Wingfield (1999), run the gamut from traditional modes of public participation (public meetings, consultation documents, committee co-options), to those which are customer oriented (complaints/suggestion schemes, satisfaction surveys), through to innovative methods designed to consult citizens on issues (focus groups, citizens' panels and referenda) or, alternatively, encourage citizen deliberation over issues (citizens' juries, community needs assessments and plans, visioning studies). The selection of any method depends on the type and depth of information required, costs and timescale, and the target audiences of frequent users, special interest groups, the general public, infrequent users, potential users and non users.

In the United States the passing in 1993 of the Government Performance and Results Act by Congress heralded a renewed interest in public consultation by agencies, not least because the law requires regular planning, measuring and reporting on results for citizens and thus encourages internal and external stakeholder input in the design and measurement of public programs. As outlined in Table 1.2 below, the legislation has several purposes:

Table 1.2 The Government Performance and Results Act, 1993

- to improve the confidence of the people in the government by holding agencies accountable for achieving program results;
- to stimulate reform with a series of pilot projects that could be used as examples for others;
- to promote a focus on results, service quality and public satisfaction;
- to help managers improve service delivery by requiring them to plan for meeting program objectives and providing them with information about program results;
- to improve congressional decision-making by providing information on achieving statutory objectives and relative effectiveness of various programs;
- to improve internal management of the federal government.

Source: Radin, B (1998) The Government Performance and Results Act: Hydra-Headed Monster or Flexible Management Tool? *Public Administration Review*, Vol 58, No 4, p308.

In each instance the language illustrated within these tables is extremely positive, and reflects a concern to respond to a more widely cited public skepticism in government. There has emerged a deficit of trust in the ability of public officials to effectively deal with the wicked problems of society and in this context citizens today are more ready to challenge perceived technocratic hegemonies built on foundations of professional knowledge and expertise. A responsiveness to service constituencies which makes use of citizen involvement, it could be hypothesized, can thus lead to a deepening relationship of mutual obligation and trust and provide an opportunity to renew credibility in public policy formation and implementation. The perceived dynamic at work, therefore, is one which combines different components of representative, consultative, direct and deliberative democracy in order to attempt to create a more open, participative and responsive polity (Pratchett, 1999).

Finally, the shift from government to governance also brings with it a need to revisit the meaning and practice of accountability. Traditionally, the concept of accountability has been interpreted as having political and administrative dimensions. The former describes the responsibility of those entrusted with the political decision-making process and involves an accountability to the electorate and state institutions such as Westminster or Congress. The latter relates to the obligation of public officials to act within the law and to apply rules and procedures in accordance with legislation. However, there is also accountability towards citizens or society as a whole and which, in turn, provides the rationale for a new approach towards relationships between public administration and citizens. Such accountability recognizes that citizens are not passive objects of the administrative process but have an active role to play (Kroger, 2001). This more direct accountability requires coherent and transparent consultation processes constructed around information, explanation and justification, but the outcomes can be an increase in the acceptability of public policy proposals, as noted above, together with a more complete demonstration of political accountability. The implications for governance systems revolve around issues of what to consult on and when, the criteria for identifying participants, practical guidance on carrying out the consultation procedure, how to present the results to decision-makers, and how to provide feedback to consultees on the contribution made by their input to the policy debate.

However, there do remain uncomfortable questions about the degree to which this re-fashioned accountability can really strengthen democratic decision-making. Burton and Duncan (1996), for example, have raised queries on who sets the agenda and who formulates the questions, the scope of the exercise and the status of the outcome. Golden (1998) has sought to address the even deeper question: whose voices get heard? In an analysis of US federal agency responsiveness to public interest groups it is noted, for example, that the agency tends "to hear most clearly the voices that support the agency's position" (p261). The danger in all this activity is that the departure point is the unstated assumption that people *ought* to participate, whereas if it does matter, there is a need to understand the relationship between accountability and participation more fully. The bleak alternative suggested by Burton and Duncan is that "attempts to change the status quo are likely to fail".

This contextual analysis of the shift from government to governance is of critical significance in this book. Governance, in short, is a process of participation which depends on networks of engagement, which attempts to embrace diversity in contemporary society, which promotes greater responsiveness to service users and, in so doing, seeks to reshape accountability relationships. The chapter now moves on to review the centrality of civil society in these governance arrangements. It is the conceptual glue that can bond government, citizens and the market within the public policy arena. Indeed, a key argument in this book is that a vibrant civil society is *the* essential characteristic of participatory governance.

Participatory governance and civil society

Public life is increasingly marked by two almost contradictory characteristics. First, there is a perceived decline of probity, integrity and honesty among those entrusted with the conduct of public affairs (Barberis, 2001). A combination of frenzied media coverage, audit office investigation and legislative committee scrutiny has conditioned a loss of trust and a deep belief in the absence of virtue. Politicians and career officials have been equally cited as transgressors and the response is a heady cocktail of louder 'spin' and ever more rules and regulations. But secondly, there is growing appreciation for the idea of civil society based on communal association, dutiful citizenship in the form of obligation to others, and partnership between the associational sector, the state and the market. The analysis in the previous section demonstrates that active participation by the public is an intrinsic feature of the public service ethos. Civil society as an organizing idea for participatory governance has the potential to confront uncertainty and disillusionment and be the engine for democratic renewal.

The concept of civil society has generated a vast literature and is subject to multiple meanings (see for example: Eberly, 2000; Seligman, 1992). In a special issue of *The Economist* Grimond (2002) leads with the provocative statement: "You'll hear it mentioned ever more often, but what is it?" He suggests that while definitional agreement can be elusive, it ultimately can be judged by the company it keeps! Expressed most simply, civil society can be recognized as the social space in which individuals are able to engage in a range of activities through informal association. In this guise it is neutral of the state and the market and has no political or ideological role. At a higher level, civil society can be defined as comprising a wide range of institutions such as voluntary organizations, faith-based communities and educational bodies. This more formal set of associations is concerned with building networks of trust and reciprocity which lie close to the heart of citizenship. It is thus inextricably linked with the concept of social capital (Baron *et al*, 2000; Fukuyama, 1995; Putnam, 1993, 2000). A third perspective, which draws on the critique of Antonio Gramsci in his *Prison Notebooks*, is that civil society is a site of ideological struggle between the dominant, hegemonic class and counter-hegemonic movements (Day *et al*, 2000). Civil society is thus an institutional form of contestation which can vary from the representation of workers rights by trade unions, to organized protest against the excesses of global capitalism. A further interpretation of civil society is linked to its perceived contribution to social cohesion and social justice. This more normative stance is readily identified with Giddens (1998, 2000) and holds that good governance depends on a constrained interaction between the state, the economy and civil society. Each component, he has argued, is a key area of power which individually needs to be tempered in the interests of social solidarity. Thus, for example, Giddens suggests that a market structure which is too dominant will bring about failure of government equity and distributional objectives, an overweening state can generate catastrophic inefficiencies, and a too aggressive civil society can threaten representative democracy. Excessive weaknesses in each element can be equally debilitating and thus the challenge is to continuously negotiate an effective social contract based on the slogan 'no rights without responsibilities':

> Those who profit from social goods should both use them responsibly, and give something back to the wider social community in return. Seen as a feature of citizenship, 'no rights without responsibilities' has to apply to politicians as well as citizens, to the rich as well as the poor, to business corporations as much as the private individual. (Giddens, 2000, p52)

The contribution of Giddens demonstrates very well that the public decision-making process is composed of three major sets of stakeholders: a state sector, a business sector and a civic or associational sector. In other words, public decision-making transcends the formal institutions of government to more comfortably rest within the sphere of governance which Healey (1997) has aptly defined as 'the management of the common affairs of political communities' (p59). It is from within this context that an even bolder interpretation of civil society can be advanced - one which holds that the arena of interactions between these three sectors comprise this phenomenon characteristic of advanced democratic societies. Cohen and Arato (1992) argue that such an arena is essential for social integration and public freedom and that, conceived in this fashion, civil society can only be matured in democratic societies characterized by an open political system and a market economy. However, at a global scale social movements and grass-roots groups can assist with the breakout from the alternative condition which Anheier (2001) labels as the 'dark side' or 'uncivil society' (p229) characterized by crime, terrorists, supremacists, and fundamentalists together with the movements, networks and organizations they create and operate.

It is also suggested in the literature that it would be misleading to identify the functions and interests of civil society with everything that takes place either within, or outside of the administrative actions of government or the activities of business. The key point here is that these behaviors reflect some interest and involvement by the general public which can express itself through a variety of alliances to include intimate associations such as the family unit, social movements and the media. The interests and actions of the civic structure are seen as influencing and mediating the structures of government and business. But because there are different and competing purposes and values between the three sectors, as well as variations in perceived and real power, there inevitably are corresponding real tensions in civil society associated with the practice of working together. More precisely this poses the question: 'Why is the planning and implementation of public projects so difficult?' An example is the inability of the many and varied urban interests (governments, businesses, environmental and citizen issue groups) to come to terms with the competing and complex expectations for planning and implementing an inter-modal transportation system in Washington DC. The outcomes of this ongoing problem, as outlined in Chapter 7 of this book, are an unsatisfactory combination of confusion, stalemate and inaction.

More generally, citizens will view such public policy confusion and apparent lethargy as negative, destructive and unnecessary behavior by their different tiers of government. On the other hand, it could be argued that conflict on public policy issues, if managed and constructively channeled, can be a positive dynamic in a pluralist society (Coser, 1964). It can energize and sustain this broader conception of civil society. But within the context of participatory governance, the challenge is to understand the complexity of the underlying interactive processes and to learn how to be effective participants in those processes. Briefly stated, how can conflict be overcome without destroying the potential for implementing any public policy initiative? The case studies and analysis in this book are concerned with these matters, all of which can be located within an interactive model of participatory governance.

Public policy decision-making in civil society: an interactive model of participatory governance

The complexity of the public policy decision-making process is illustrated in Figure 1.1 which seeks to portray the interactions of participatory governance across the trinity of government, the associational sector and the business sector.[2] As argued above, the relational spaces created by these three dimensions of contemporary, advanced capitalist democracies help locate civil society. In reading the diagram from top to bottom it is the case that each sector comprises multiple actors. Thus government contains the political executive arena, an elected arena, an appointed arena and a career officer arena; the business sector contains customers and consumers, enterprises, regulatory mechanisms and financial markets; and the associational sector is made up of individuals, micro and macro special interest groups, faith-based groups and community groups. Tighter relationships as dictated by electoral politics and public accountability among government actors, as well as by the profit motive among business actors are in contrast with looser connections among actors in the associational sector. At policy formation, operational and technical levels the interactions are a combination of consultation, lobbying and resource sharing at varying degrees of intensity. These activities reflect different perspectives on primary aspirations which for the government sector is responding to the public will, for the associational sector is enhancing community well-being,

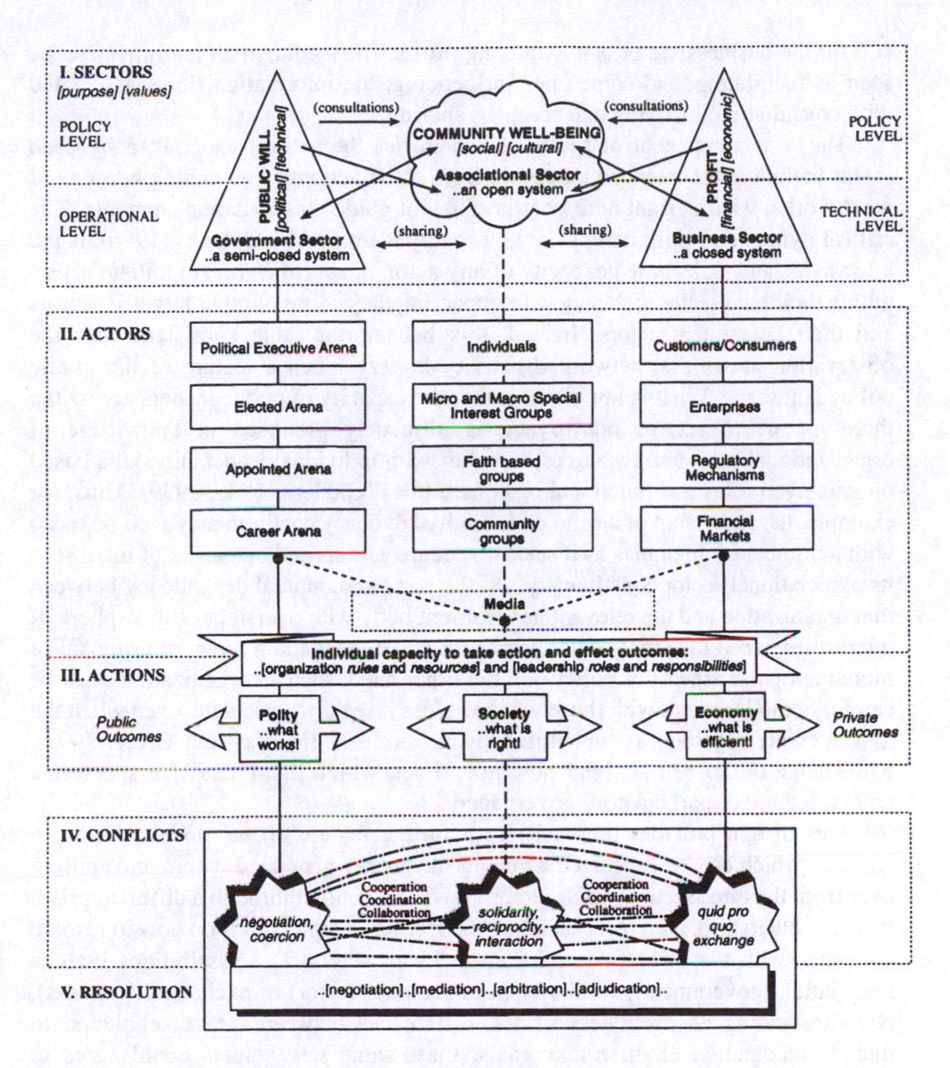

Figure 1.1 The Civil Society: An Interactive Model of Participatory Governance

and for the business sector is maximizing profit. The media, in all its forms, can be seen as facilitating, and sometimes influencing, the information flows associated with consultation, lobbying and resource sharing.

The next component of the model identifies the actions undertaken by each sector including the intended outcomes from those actions. The homing beacons of what works, what is right and what is efficient guide these accomplishments. The critical dynamic at work here, as more broadly identified by Giddens (1986), is the knowledge and skills based capacity of any actor, or sets of actors, to initiate action within their own sector with a view to impacting the actions of other targeted sectors and their constituent actors. Indeed, this behavior is quite consistent with the observation above that networked activity is very much a feature of the public policy landscape. What is important is that the capacity of public leaders across the three sectors to act, or not to act, is ultimately grounded in knowledge of organizational rules and resources, together with individual leadership skills based on perceived roles and functional responsibilities (Giddens, 1986, 1999). Thus, for example, the leadership of a rural community advocacy coalition may well be faced with a number of dilemmas as it seeks to secure actions and outcomes of interest to its associational sector constituency. On the one hand, mutual dependency between that organization and the relevant government body with oversight of this sphere of interest may lead towards greater enclosure and a dampening down of criticism of more unpopular aspects of policy. On the other hand, such an organization must be careful not to lose faith with those whose voices it seeks to articulate, even when the messages are regarded as 'unhelpful' by government (Murray and Greer, 1999). Knowledge on how to act, and how not to act, within this interactive space is a crucial feature of participatory governance.

This insight provides the basis for a further feature of the model relating to conflict, which can be conceived as being the energy generated within and spilling over from the three sectors. Each sector views any conflict through a different prism from the others and, therefore, each holds differing interpretations on how to resolve an issue with the other two and their relevant actors. The possibilities include negotiation (government), solidarity (associational sector) or exchange (business). Notwithstanding an awareness of the differences between the stakeholders, to include recognition of dissimilar goals, these same stakeholders could agree to manage their conflict through cooperation, coordination or collaboration. These terms are frequently interchanged, but each has its own subtleties which reflect variations in mutual relationships and goals, jointly developed structures and shared responsibility, mutual authority and accountability for success, and the sharing of resources and rewards (Mattessich and Monsey, 1992). In short, collaboration represents a desire to work together by sharing resources to achieve a common goal. Coordination represents a desire to partner, but not share resources, to achieve a common goal. Cooperation represents a sharing of ideas but neither work nor resources (Hughes *et al*, 1998). While collaboration under these criteria is often placed at the high end of the partnership continuum, it is perhaps more appropriate to consider the application of each option as being context specific. Again this reiterates the importance of knowledge on how to act in maintaining often fragile coalitions.

But finally, the working through of an appropriate arrangement may well require the adoption of more overt resolution oriented/conflict management techniques involving negotiation, mediation, arbitration or adjudication. In this sense, conflict should not be perceived as a destructive or negative process (Coser, 1964). It can be a positive factor in sustaining a healthy civil society and, thus, the challenges for the practitioner are to understand the complexities of public policy decision-making and to search for effectiveness of effort in the process of participatory governance. The case studies contained in this book provide valuable insights on these matters. Essentially, therefore, the model above can be summarized as a process in which stakeholders with different interests and objectives arrive at a shared definition of a problem and an agreed set of policy measures through negotiations and the exchange of resources. However, as Schedler and Glastra (2001) go on to point out, there is a need to be aware of a number of serious constraints which can have a bearing on this form of interactive policy making. They identify several difficulties: professional resistance to the adjustment of long established institutional competencies constructed on the basis of training and experience; the competing and different values of sectoral participants which can clash during the search for common ground; the uncertainty about how agreement will be received by sectoral constituencies; the inclusiveness of the process to the extent that everyone who should be invited is given an opportunity to participate and the subsequent representativeness of these actors; unequal relationships on the basis of status, access to data, capacity to engage in the discussions, negotiating experience and strategic insight; and finally, the degree to which participants produce a weak compromise or lose sight of external, serious realities as a result of comfortable association. In other words, the interactive public decision-making process in civil society is not risk free which again highlights the importance of reflective participatory practice.

At a broader level, there are, of course, a number of prominent dialectics in the relationship between participatory governance and democratic government. The next section of this chapter deals briefly with these issues which should be regarded as an additional 'reality check' for the now much vaunted rhetoric of participation.

Key issues in participatory governance

It would be entirely misleading if the analysis above was to ignore the reality that participatory governance is not without its conceptual difficulties and operational constraints. The contribution of multiple actors to public policy decision-making can be incisively rewarding or disappointing and, as Thomas (1995) has argued, much depends on the understanding of the public manager about when and how to invite that involvement:

> When successful, public participation can bring substantial benefits - more effective public decisions, a satisfied and supportive public, and most important, a stronger democracy; but when it fails, and it has frequently failed, public participation can leave in its wake a dissatisfied and even restive public, ineffectual decisions, and a weakened if not faltering democracy. The risks of failure have too often persuaded public managers to avoid or minimize public involvement. No choice could be more foolhardy. Public involvement, though neither for all matters nor always to the same extent, is now essential for effective public participation. (p2)

The unpacking of this apparent dilemma, especially when set against the civil society model in the previous section of this chapter, requires consideration of the following inter-linked questions: (1) are participatory instruments merely functional procedures for energizing governance? (2) is emphasis on the contribution of the associational sector opportunistic? and, (3) what are the barriers to authentic participation?

In relation to the first issue, there appears to be growing pressure over the past decade to introduce participatory and direct forms of democracy into government in order to revitalize representative democracy. Within the United Kingdom, for example, public participation has been more specifically linked with the current emphasis by Government on *democratic renewal* - a term which Pratchett (1999) identifies as capable of being interpreted in three ways: (1) as a set of practical solutions to electoral apathy, together with functional impotence and arcane decision-making structures in local government; (2) as a systemic failing in the attitudes, beliefs and behavior of citizens in relation to democracy which has conditioned disinterest in traditional modes of political participation, mistrust of the public sector and an indifference to the rights and responsibilities of citizenship; and (3) as a normative expression of a new mode of democracy in which different components of representative, consultative, direct and deliberative democracy are integrated to create a more open and responsive polity. The debate is not whether one form of democracy is better than another, but rather how best to harmonize these different approaches with their intrinsic advantages.

Thus, within the United Kingdom, policy documents such as *Modern Local Government: In Touch with the People* (July 1998) and *Modernizing Government* (March 1999) have emphasized the pivotal role ascribed to the general public in determining service provision. This is recognition that participatory democracy works better than representative democracy in revealing preferences (Walsh, 1996). Pratchett's views on the contribution of deliberative democracy to democratic renewal are especially pertinent here:

> ... the value of deliberation is in the actual participation. Activities such as 'visioning' can be very effective in building consensus across communities. They can also serve as a means of 'educating' citizens in civic values by providing valuable experience of political participation. Consequently, deliberative participation ... serves two potential functions: the 'education' of citizens and the 'transformation' of views through discussion. (p12)

But Pratchett, in line with Thomas (1995), also draws attention to some of the potential difficulties in the deliberative approach. The emphasizing of community differences, increased parochialism, and the raising of unrealistic expectations of what can be achieved require careful management in order to avoid even greater disillusionment. Accordingly, democratic renewal requires effort at rebuilding trust across central - local relations. At the heart of that challenge must lie, as argued above, the task of nurturing a strong civil society to connect individuals in their places and associations with government and the market. The thrust of this analysis would suggest, therefore, that this shared involvement is central to energizing *creative* governance.

A second major issue is the extent to which the emphasis on the associational sector in relation to participatory governance is opportunistic exploitation by public managers. This is rooted in the apparent distance being opened up between representative institutions and citizens through what has been dubbed 'the hollowing out of the state'. The phenomenon is associated *inter alia* with the contracting-out or devolution of activities traditionally undertaken by government to non- or quasi-governmental organizations, further deepening the network structure of contemporary society (Howlett, 2000). The associational sector is actively involved in service provision and in developed democracies receives considerable support from government in the form of grants, contracts or service agreements. In transitional countries, by way of contrast, the associational sector is more a substitute for full public sector provision and, so, joint ventures in social services can increase availability beyond what would be possible under purely public or private regimes (Bucek and Smith, 2000). In order for this to happen, organizations and movements in the associational sector must be voluntary, self-developed and self-mobilized. They must be free of direct control by the government sector and exist on the basis of their capacity to achieve and maintain organizational status. As such they are generally recognized as having the potential to operate within and impact on society. These organizations have legal rights (both political and property rights) and are not intended to be constrained by civil law (Cohen and Arato 1992).

Legislation on occasions has sought to deepen opportunities for the associational sector in these activities and thus, for example, in the United States the content of the 1996 Personal Responsibility and Work Opportunity Reconciliation Act includes a 'Charitable Choice' requirement that states contract with faith based social service providers on the same basis that they contract with other non-profits (Kennedy and Bielefeld, 2002). At the same time it is important, indeed critical, that the associational sector is independent and commands status through a detachment from central government, even if many do receive funding from the state. The associational sector should not, therefore, be viewed as supine and again, in the context of the United States, there is evidence to suggest that many faith-related agencies fear the consequences of further entanglements with government and do not wish to take on the responsibilities of the welfare state (Smith and Sosin, 2001). The governance role of the associational sector, therefore, is both differentiated and contested.

The role of the associational sector can take many forms. It is concerned with service delivery, but it is also committed to exerting pressure for change, being an advocate for service users, and articulating the needs of those whose voices are not readily heard by government. The challenge for public managers in dealing with claims of opportunistic exploitation is to develop genuine negotiation and bargaining strategies related to resource sharing as a follow on to the more basic consultation and lobbying activities which characterize many government - associational sector interactions. In the United Kingdom the development in 1998 of a *Compact* of recognition principles and mutual undertakings between government and representatives of the associational sector opens the door to mainstreaming the latter within the public policy agenda (Kendall, 2000) and provides a potentially useful illustration of appropriate policy entrepreneurship.

The final and crucial issue to be highlighted at this stage is part of a more fundamental critique of public involvement which asks questions around participation authenticity. Certainly the claim that the discernible growth in participatory initiatives enhances democratic renewal is one that needs to be treated with caution (Lowndes *et al*, 2001). Their research on local government in England indicates that the declared goals of empowering citizens or increasing their awareness were of secondary importance to the more tangible benefits of improving decision-making, notwithstanding some evidence that the introduction of public participation did slow down that decision-making process. Linked with this are other concerns about the capturing of issues by particular groups who are not representative of the wider community, the tendency to concentrate on relatively trivial issues, the danger of consultation overload and the possibility of groups losing confidence in the decision-makers if participatory initiatives are not perceived as being successful. Meadowcroft (2001) expresses similar concerns about the limitations of deliberative 'thickness' in helping to create a more participatory democracy:

> While a political system that cannot change will not endure, these unreal expectations will only lead to disappointment and disillusionment, creating a momentum for more thorough and profound measures, which will only fail more thoroughly and more profoundly. (p40)

All these concerns bolster the case for a continuation of the search for authentic participation and not surprisingly there is no shortage of analysis and advice related to this quest. As argued by Weeks (2000), if deliberative democracy aspired only to the narrow goal of informing policy makers of the judgment of citizens, participation could be minimized to a small statistically representative sample. Its ambitions are, however, much grander and encompass the revitalization of civic culture, improving the nature of public discourse and generating 'the political will necessary to take effective action on pressing problems'. What is clear is that those who hold power in relation to decision-making, control of resources and implementation must be prepared to share that power with other stakeholders; actors located in the business and associational sectors must be willing to engage and believe that there is value in so doing (Foley and Martin, 2000). Different participation strategies and methods are necessary to reach different citizen groups; high conflict issues may require early citizen involvement to encourage consensus. This may include the development of alternatives in order for the public to accept the final outcome thus helping to legitimize the policy process (Walters *et al*, 2000). In this vein Thomas (1995) goes as far to suggest that more public participation is appropriate when the acceptance of a decision is important, and less public participation is appropriate when the quality of a decision is important. And finally, there is a need to redefine the role of expertise in public administration and to change the training curriculum by including communication, listening, team building, meeting facilitation and self knowledge skills (Simrell King *et al*, 1998).

Conclusion: participatory governance through interactive public decision-making

This chapter has explored the underlying dynamics of a broadly recognized shift from government to governance and has demonstrated that public decision-making has now become an interactive process involving multiple actors from the government, business and associational sectors. Each brings information and resources to networks of interest which under ideal circumstances can secure shared definitions of and interdependent responses to problem conditions. Interactive public decision-making can be initiated by any of the three sectors, but will subsequently require action oriented outcomes constructed around relationships of cooperation, coordination or collaboration. It is vital that those entering any arena of negotiated policy making understand that real difficulties can lie ahead linked to professional resistance, competing values, hostile followers, unequal relationships of power, controlled participation and suffocation by comfortable association. Blockages connected with the deliberative processes may require the introduction of conflict resolution strategies to get everyone to 'yes'. In the end, however, the quality of this public decision-making in civil society will be dependent on the degree to which the stakeholders accept their involvement as a shared responsibility. Herein lies the deepest truth of participatory governance!

The chapters in the next two sections of this book offer a range of narratives consistent with this theme of interactive public decision-making in civil society. The essential distinction underpinning their arrangement is that the book should progress from institutional perspectives towards operational concerns. The laboratories of analysis span the United States, Europe, Africa and Australia. The book concludes with a cross-cutting synthesis of this material and seeks not only to respond to many of the questions raised by this introductory chapter, but also to identify insights for participatory governance policy and practice. Stated as principles, these are derived directly from the previous chapters, but deserve to command wider application.

References

Anheier, H. (2001) 'Measuring global civil society', in Anheier, H., Glasius, M., Kaldor, M. (eds) *Global civil society*, Oxford: Oxford University Press.

Barberis, P. (2001) 'Civil society, virtue, trust: implications for the public service ethos in the age of modernity', *Public Policy and Administration*, Vol. 16, No. 3, pp. 111-126.

Baron, S., Field, J. and Schuller, T. (2000) *Social capital: critical perspectives*, Oxford: Oxford University Press.

Bekke, H., Kickert, W. and Kooiman, J. (1995) 'Public management and governance', in Kickert, W. and van Vught, F. (eds) *Public policy and administrative sciences in the Netherland*, London: Harvester Wheatsheaf, pp. 201-218.

Bryson, J. and Crosby, B. (1992) *Leadership for the common good: tackling public problems in a shared world*, Jossey Bass Publishers, San Francisco.

Bucek, J. and Smith, B. (2000) 'New approaches to local democracy: direct democracy, participation and the 'third sector'', *Environment and Planning C: Government and Policy*, Vol. 18, No. 1, pp. 3-16.

Burton, P. and Duncan, S. (1996) 'Democracy and accountability in public bodies: new agendas in British governance', *Policy and Politics*, Vol. 24, No. 1, pp. 5-15.

Cloke, P., Milbourne, P. and Widdowfield, R. (2000) 'Partnership and policy networks in rural local governance: homelessness in Taunton', *Public Administration*, Vol. 78, No. 1, pp. 111-133.

Cohen, J. and Arato, A. (1992) *Civil society and political theory*, Cambridge: MIT Press.

Coser, L. (1964) *The functions of social conflict*, New York: The Free Press.

Day, G., Dunkerley, D. and Thompson, A. (2000) 'Evaluating the 'New Politics': civil society and the National Assembly for Wales', *Public Policy and Administration*, Vol. 15, No. 2, pp. 25-37.

Department of the Environment, Transport and the Regions (1998) *Modern local government: in touch with the people*, London.

Eberly D. (ed) (2000) *The essential civil society reader: the classic essays*, Oxford: Rowman and Littlefield Publishers.

Foley, P. and Martin, S. (2000) 'A new deal for the community? Public participation in regeneration and local service delivery', *Policy and Politics*, Vol. 28, No. 4, pp. 479-491.

Fukuyama, F. (1995) *Trust: the social virtues and the creation of prosperity*, London: Hamish Hamilton Ltd.

Giddens, A. (1986) *The constitution of society*, Berkeley: University of California Press.

Giddens, A. (1998) *The third way: the renewal of social democracy*, Cambridge: Polity Press.

Giddens, A. (1999) 'Focus of the third sector in the Third Way', *Paper No 2*, West Malling: UK Charities Aid Foundation.

Giddens, A. (2000) *The third way and its critics*, Cambridge: Polity Press.

Golden, M. (1998) 'Interest groups in the rule making process: who participates? Whose voices get heard?' *Journal of Public Administration Research and Theory*, Vol. 8, No. 2, pp. 245-270.

Goodwin, M. (1998) 'The governance of rural areas: some emerging research issues and agendas', *Journal of Rural Studies*, Vol. 14, No. 1, pp. 5-12.

Gould, C. (1996) 'Diversity and democracy: representing differences', in Benhabib, S. (ed) *Democracy and difference: contesting the boundaries of the political*, Princeton: Princeton University Press, pp. 171-186.

Gramberger, M. (2001) *Citizens as partners: OECD handbook on information, consultation and public participation in policy-making*, Paris: OECD.

Grimond, J. (2002) 'Civil society', *The Economist - Special Issue*, p. 18.

Hansen, K. (2001) 'Local councilors: between local government and local governance', *Public Administration*, Vol. 79, No. 1, pp.105-124.

Healey, P. (1996) 'Consensus-building across difficult divisions: new approaches to collaborative strategy making', *Planning Practice and Research*, Vol. 11, No. 2, pp. 207-216.

Healey, P. (1997) *Collaborative planning: shaping places in fragmented societies*, London: Macmillan.

Howlett, M. (2000) 'Managing the 'hollow state': procedural policy instruments and modern governance', *Canadian Public Administration*, Vol. 43, No. 4, pp. 412-431.

Hughes, J., Knox, C., Murray, M. and Greer, J. (1998) *Partnership governance: the path to peace in Northern Ireland*, Dublin: Oak Tree Press.

Kendall, J. (2000) 'The mainstreaming of the third sector into public policy in England in the late 1990s: whys and wherefores', *Policy and Politics*, Vol. 28, No. 4, pp.541-562.

Kennedy, S. and Bielefeld, W. (2002) 'Government shekels without government shackles? The administrative challenges of Charitable Choice', *Public Administration Review*, Vol. 62, No. 1, pp. 4-11.

Kettl, D. (1999) comments cited in Roberts, A. 'Reinventing government: REGO around the world', *Government Executive Magazine,* January.

Kroger, M. (2001) 'Report of Working Group "Consultation and participation of civil society"', *White Paper on European Governance: Work Area No 2 - Handling the Process of Producing and Implementing Community Rules,* Brussels: European Commission.

Leach, S. and Wingfield, M. (1999) 'Public participation and the democratic renewal agenda: prioritization or marginalization?', *Local Government Studies*, Vol. 25, No. 4, pp. 46-59.

Lowndes, V., Pratchett, L. and Stoker, G. (2001) 'Trends in public participation: part 1 - local government perspectives', *Public Administration*, Vol. 79, No. 1, pp. 205-222.

MacGreil, M. (1996) *Prejudice in Ireland revisited*, Maynooth: National University of Ireland.

Mattessich, P. and Monsey, B. (1992) *Collaboration: what makes it work*, Saint Paul: Amherst H Wilder Foundation.

Meadowcroft, J. (2001) 'Community politics, representation and the limits of deliberative democracy', *Local Government Studies*, Vol. 27, No. 3, pp. 25-42.

Modernising government (1999) White Paper presented to Parliament by the Prime Minister and the Minister of the Cabinet Office, London: The Stationery Office.

Murray, M. and Greer, J. (1999) 'The changing governance of rural development: state-community interaction in Northern Ireland', *Policy Studies*, Vol. 20, No. 1, pp. 37-50.

Osborne, D. and Gaebler, T. (1992) *Reinventing government: how the entrepreneurial spirit is transforming the public sector*, New York: Plume.

Peters, B.G. and Pierre, J. (1998) 'Governance without government? Rethinking public administration', *Journal of Public administration Research and Theory*, Vol. 8, No. 2, pp. 223-243.

Pratchett, L. (1999) 'Introduction: defining democratic renewal', *Local Government Studies*, Vol. 25, No. 4, pp. 1-18.

Putnam, R. (1993) *Making democracy work: civic traditions in modern Italy*, Princeton: Princeton University Press.

Putnam, R. (2000) *Bowling alone: the collapse and revival of American community*, New York: Simon and Schuster.

Radin, B. (1998) 'The Government Performance and Results Act (GPRA): Hydra-headed monster or flexible management tool?', *Public Administration Review*, Vol. 58, No. 4, pp. 307-316.

Rhodes, R. (1997) *Understanding governance: policy networks, governance, reflexivity and accountability*, Buckingham: Open University Press.

Roberts, A. (1999) 'Reinventing government: REGO around the world', *Government Executive Magazine*, January.

Schedler, P. and Glastra, F. (2001) 'Communicating policy in late modern society: on the boundaries of interactive policy making', *Policy and Politics*, Vol. 29, No. 3, pp. 337-349.

Seligman, A. (1992) *The idea of civil society*, Princeton: Princeton University Press.

Simrell King, C., Feltey, K. and O'Neill Susel, B. (1998) 'The question of participation: towards authentic public participation in public administration', *Public Administration Review*, Vol. 58, No. 4, pp. 317-326.

Smith, S., Rathgeb and Sosin, M. (2001) 'The varieties of faith related agencies', *Public Administration Review*, Vol. 61, No. 6, pp. 651-670.

Thomas, J. (1995) *Public participation in public decisions: new skills and strategies for public managers*, San Francisco: Jossey-Bass Publishers.

Thompson, J.R. and Ingraham, P.W. (1996) 'The reinvention game', *Public Administration Review*, Vol. 56, No. 3, pp. 291-297.

UK Government Cabinet Office (1998) *An introductory guide: how to consult your users*, London: Cabinet Office.

UK Government Cabinet Office (1999) *Involving users: improving the delivery of local public services*, London: Cabinet Office.

UK Government Cabinet Office (2000) *Code of practice on written consultations*, London: Cabinet Office.

Walsh, K. (1996) 'Public services, efficiency and local democracy', in King, D. and Stoker, G. (eds) *Rethinking local democracy*, London: Macmillan, pp67-88.

Walters, L., Aydelotte, J. and Miller, J. (2000) 'Putting more public in policy analysis', *Public Administration Review*, Vol. 60, No. 4, pp. 349-359.

Weeks, E. (2000) 'The practice of deliberative democracy: results from four large scale trials', *Public Administration Review*, Vol. 60, No. 4, pp. 360-372.

Notes

1 The associational sector includes what is frequently labeled the 'third sector' of organized voluntary and community groups.

2 This civil society model was developed during the course of W. Robert Lovan's doctoral studies at the Institute for Conflict Analysis and Resolution, George Mason University, Virginia, USA.

Section Two:
Institutional Perspectives on Participatory Governance

Chapter 2

Evolving Participatory Governance and Developmental Local Government in Post-Apartheid South Africa

Etienne Nel

Introduction

The 1990s witnessed the remarkable transformation of South African society and government from one of the more repressive to one of the most liberal on earth. Though the shift from ruthless, top-down apartheid ideology and state management to a democratic society and state has not been without its difficulties, it nonetheless remains one of the most dramatic and significant political and social transitions of the modern period (Waldmeir, 1997). The African National Congress (ANC)-led government is now striving to redress the widespread and deeply entrenched imbalances which represent the legacy of decades of apartheid policies. In planning for development in one of the most unequal societies in the world, the South African government has decided to strengthen both grassroots participation and delivery by placing considerable emphasis on what it terms 'developmental local government'. The government argues that the locus for such activity should be local government and that, 'the central responsibility of municipalities (is) to work together with local communities to find sustainable ways to meet their needs and improve the quality of their lives' (RSA, 1998a, p17). In order to achieve this objective of 'developmental local government', local authorities are now expected to maximize both social development and economic growth and to work with communities to ensure that local economic and social conditions are conducive for the creation of employment opportunities (Nel and Binns, 2001). The implicit bond between participatory governance and development is clearly a key theme in South African policy and planning. Given the priority now being accorded to local government as the venue for participation, this chapter will largely focus on local governments, participation within them and associated development strategies.

The new South African Constitution, widely regarded as the most liberal in the world, has entrenched a noteworthy range of civil liberties and freedoms. Together with supporting legislation it enshrines the concept of 'participatory governance' (RSA, 2000a) and ensures the right of every citizen to participate in the democratic decision-making process, particularly at the level of local government. Though sound in principle, there are unfortunately often practical difficulties which impede the application of such rights. Despite this, the concept of popular participation in local-level governance cannot be divorced from current, parallel trends, to decentralize power to the lowest levels of government administration and, simultaneously, to enhance the responsibilities of local government to encourage community participation in order to fulfill its new mandate of 'developmental local government' (RSA, 1998a). So significant has this transition been that the South African government argues that it has shifted administration from a hierarchical or 'tiered' chain of operation i.e. from one operating in a top-down fashion, to one in which three parallel 'spheres' of government - national, provincial and local - enjoy equal status on a hierarchical plane (RSA, 1996a; 1998a).

Whereas political change was the hallmark of the first half of the 1990s in South Africa, it was only in 1996 that a new national constitution, which enshrines the principles of popular participation, was promulgated. This has set in motion a process of change in the laws and operating procedures of local government, which only led to the formalization of a new local government dispensation in 2000. Hence, although high levels of civil society participation in decision-making have been a characteristic of development in the country for several years, it was only in 2000 that the final parameters were put in place to legally define and encourage such participation at the local government level. South Africa's second democratic local government elections, which were held on 5 December 2000, ended the 'transition' phase of post-apartheid local government (1995-2000) and initiated a new era in municipal administration. In addition to encouraging popular participation, as will be detailed below, local authorities are now expected and have been empowered to play an increasingly prominent role in local economic development and employment creation (RSA, 1998a; 2000a). Government has committed itself to an agenda which, in general terms, upholds the principles of equity, diversity and interdependence through a commitment, in principle, to promote development, racial unity and participation by all citizens in governance. This is reflected in the country's national motto, affixed to the national coat of arms. Based on the ancient Khoisan language of one of the oldest societies on earth, which pre-dates all current dominant racial and ethnic groups, the motto reads: '!ke e:/xara //ke' which broadly translates as 'unity in diversity' or 'diverse people unite'. The central feature of the coat of arms supports this theme, depicting two human figures greeting each-other in a symbol of 'greeting and unity' (RSA, 2001, http://www.polity.org.za/govdocs/misc/cofarms00. html 21 October 2001). These symbols of state reflect a commitment to a new dispensation and social order. South Africa, therefore, provides a very interesting example of a nation striving to come to terms with its history and diverse cultures through the promotion of mechanisms to ensure equality, participation, development and interdependence.

Though differing in focus and outcome, what is happening in South Africa broadly parallels what is clearly an emerging theme around the world, namely that, over the last 15 years an influential literature has emerged about the rise of a broader and more fragmented local governance, extending beyond the formal agencies of local government to include a 'wide range of other actors, institutional and individual, private and voluntary and public sector, which are involved in regulating a local economy and society' (Valler, Wood and North, 2000, p.409). However, unlike in Great Britain for example, the independence, powers, control and decision-making of local government have been significantly enhanced in South Africa. It is through the operation of strengthened local government that enhanced popular participation is expected to find expression. At this stage, although there is support for the concept of partnership formation and collaboration with other sectors, the key compact is envisaged as being between the associational sector and locally elected representatives. This chapter seeks to briefly examine the principles of participation, with particular reference to local government in Africa, before providing an outline of South Africa's bitter apartheid history which furnishes a backdrop for recent change. The core focus of this chapter comprises a critique of evolving local government policy and law on which the key mechanisms for the associated concepts of participatory governance and development are envisaged as being grounded.

Popular participation and Africa

A key theme internationally with respect to both government policy and practice and within academic debate from the 1980s has been the emphasis placed on 'community involvement in the construction and delivery of urban policy' (Raco, 2000, p573). Not only are enhanced levels of popular participation seen as essential to promote social justice (Haughton, 1998), empowerment (Lyons and Smuts, 1999) and democratic governance, but more critically it has been argued that such emphasis cannot be divorced from the wider 'neo-liberal objective of creating active citizens to promote self-reliance, local initiative and reduced "dependence" on the welfare state' (Raco, 2000, p574). Community empowerment and self-expression have been manifested through enhanced public debate and local-level politicization, and also through the emergence and participation by a wide and diverse range of community agencies, Non Governmental Organisations (NGOs) and local coalitions in matters of local-level governance. The role of such movements, particularly in coping with the challenges of cities in the developing world was endorsed by the United Nations at its Habitat II conference in 1996 (Lyons and Smuts, 1999).

The concept of participatory governance is starting to receive widespread support in Africa. According to Gooneratne and Mbilinyi (1992), in the light of the bleak economic and political prospects which face the African continent, issues of popular decision-making, decentralization and self-reliance are a key feature of the modern African state. In 1998, the United Nations Center for Human Settlements (Habitat), which is based in Nairobi, Kenya, produced a key document titled, *Towards a World Charter of Local Self-Government*, which reaffirmed the decision of the Council of European Municipalities in Versailles in 1953 to adopt the Charter of Municipal Liberties. This reflected a commitment to 'strong local institutions enjoying a high degree of democratic autonomy' (UNCHS, 1998, p4). The proposed World Charter carries forward the decisions of the Habitat II conference and seeks to enhance the principles and practice of decentralization, democracy, subsidiarity and the strengthened capacity of local government. Popular participation and civic engagement in decision-making are encouraged, as is the mandate to develop partnerships with all actors of civil society, particularly non-governmental organizations, community-based organizations and the private sector (UNCHS, 1998). These principles have received widespread support in Africa and were endorsed by representatives of African local governments who, in 2000, produced a statement that 'underlined the importance of decentralization to harnessing the creative energy of people for economic growth and development on the continent and to the process of democratization in Africa' (UNCHS, 2001, Press Release, CHS/00/, 31 March 2001, p1).

South Africa's legacy

Historically, apartheid and the governments which systematically enforced it in pre-1994 South Africa disempowered the nation's people, denied access for most of them to direct participation in any form of governance and vested all power and decision-making in a rigid, minority controlled, top-down government which acted in a hierarchical manner. Local government depended directly on central government for direction. Apartheid was a form of direct and deliberate racial discrimination which denied opportunities and representation to all people who were not white, confined many to a servile status and rigidly regulated their way of life and places of residence (Lemon, 1987; Lester et al, 2000).

The demise of apartheid in the early1990s did not eliminate centuries of entrenched inequality. In fact, many indicators suggest that while the life prospects and opportunities of middle-class black people have improved dramatically, for the impoverished majority, economic marginalization has actually increased, creating a desperate situation for which little prospect for short-term resolution appears to exist (Nel, 2000). South Africa is characterized by inherited levels of social and spatial inequality which are only exceeded by those of Brazil (Economist, 2001, p3). Examining the different groups within South Africa's population, Whites in 1993 had a Human Development Index of 0.901, not very different from that of many Western European countries. However, the HDI for the Indian population was 0.836, for Coloreds (i.e. people of mixed race) 0.663 and for Blacks only 0.5, the latter being comparable with countries such as Papua New Guinea and Cameroon (Lester, Nel and Binns, 2000). Stark spatial inequalities are also still apparent within South Africa, such that, 'sophisticated urban centers with their established business corporations, which are indistinguishable from those in the Western world, can be viewed from the same spot as the crime-ridden squatter camps, typical of those in South American cities' (Lester, Nel and Binns, 2000, p235). On a broader regional basis, those provinces which have incorporated the former 'black homelands' (i.e.

the largely rural, Black racial reserves), are generally much poorer than others. Eastern Cape province, for example, which in 1994 incorporated the homelands of Transkei and Ciskei, had the highest 'official' unemployment rate, the highest infant mortality rate and the second lowest life expectancy in South Africa in 1996 (Binns, 1998; DBSA, 2000; StatsSA, 1999).

Prior to 1994, people who were not white were effectively denied democratic representation and any legitimate means of participating in development activities. In addition, the approach to local government administration and development during the apartheid era provides a sharp contrast to that in the contemporary period. Before 1994, local government, 'was responsible for a narrow range of traditional local government functions - providing basic municipal services such as water, electricity, internal roads, street lighting, storm-water drainage, sewerage etc. - and played a minimal developmental or redistributive role' (Pycroft, 1998, p155). Local government, with certain exceptions, was largely the domain of the white minority in terms of voting rights and decision-making. Even for this group, local government powers were severely restricted, decisions were made from above and it would be difficult to argue that any significant public participation was practiced. In fact there is evidence of the definite loss of power and autonomy at the local-level in the twentieth century as apartheid reached its zenith (Nel, 1999). Solomon has argued that the approach adopted during the apartheid era was paternalistic, non-democratic and non-consultative (Solomon, 1990). Pre-1994 planning was top-down, and at the local level this was implemented on a racially segregated basis which largely reflected the needs of the privileged white minority. The dominant emphasis was on sectoral planning and infrastructural delivery programs by the public sector, with little involvement of the private sector. Issues such as environmental sustainability, economic viability, poverty alleviation and welfare were rarely considered. The apartheid regime had a rigidly hierarchical government and planning structure, such that local municipalities were in a subservient position to provincial government and then the national authority, and their plans and budgets generally had to be approved by provincial administrations. The provincial and national governments, 'exerted considerable indirect control over local planning through a dense web of racially based legislation' (DPLG, 2000, p14).

South Africa's transition and the enhancement of participation

With the demise of apartheid a radical mind-set and policy shift has taken place in South Africa, not only in terms of the axing of racial-policies and practice, but also defined moves to ensure equality and participation. The role and place of the citizen has been significantly enhanced and all forms of discrimination disallowed. This has been facilitated and enabled by the enactment of an extremely liberal constitution and the passage of a range of laws which have enhanced the role and place of local government as the most basic form of democratic participation and expression. The process is clearly still in its formative stages. As a result, in 'the post-apartheid era South Africa has seen a continued evolution of the country's democratic institutions, and the nature of South African democracy continues to develop with them' (Lyons and Smuts, 1999, p2152). Although many of the changes detailed below are innovative, striking and dramatic, it would be naive to suggest that the process has been problem free or has achieved all its goals. It would also be a major oversight to suggest that all people have benefited equally. In a country plagued by some of the world's highest crime levels, where corruption is acute and many still lack access to basic services, being aware of, and exercising one's rights is not always a straightforward process (Nel, 2000). In addition, the danger exists that officials often merely pay lip service to the principles which they are obliged to uphold (Lyons and Smuts, 1999).

On the positive side, despite these very real obstacles, the principles of participatory governance are now firmly entrenched in local government operations as is the obligation to pursue a consensual basis for project development and management which is widely adhered to in undertaking development ranging from service and housing provision to job-creation schemes. In the remainder of this section, the key policy and legal documents sketching out and facilitating both popular participation and developmental local government are outlined.

The Reconstruction and Development Program (RDP)

The RDP was launched as the key ANC policy document before the democratic elections in April 1994, and then formalized in September 1994 as the new government's 'White Paper on Reconstruction and Development'. It was designed to provide a broad framework for South Africa's new development vision, priorities and operational procedures and it aimed to both lay a basis for subsequent laws and actions to address the extreme social and spatial inequalities engendered by years of apartheid policies and to promote overall development (ANC, 1994). In a radical break with the past, the RDP was promoted essentially as a 'people-driven process', focusing, 'on our people's most immediate needs, and (relying), in turn, on their energies to drive the process of meeting these needs' (ANC, 1994, p5). The RDP also placed considerable emphasis on grassroots empowerment, suggesting that, 'development is not about the delivery of goods to a passive citizenry, but rather it is about active involvement and growing empowerment' (ANC, 1994, p5) which integrates 'all levels of the state together with non-governmental organizations and community-based organizations' (Lyons and Smuts, 1999, p2155). This emphasizes the fundamental links between participation and development which are conceptualized in South Africa.

The ANC government argued that the broad goals of the RDP could be achieved by giving much more responsibility for development to local government, which is viewed as the primary level of democratic representation. As the RDP stresses, 'the democratic government will reduce the burden of implementation which falls upon its shoulders through the appropriate allocation of powers and responsibilities to lower levels of government, and through the active involvement of organizations of civil society' (ANC, 1994, p140). Whilst these are understandable and admirable objectives, they have placed a considerable burden of responsibility on the local tier of government, a situation aggravated by very real capacity and financial constraints experienced by most of the smaller local authorities (Nel and Binns, 2001). The key features of the RDP are firstly that it laid a policy basis for enhancing participation through local government and secondly, it provided principled support for grass-roots action. As subsequent events have shown, the former has been a far more defined focus of attention.

The 1996 Constitution

The key basis for all legislation is the national Constitution of the country, which is the supreme law upon which all other laws are based (RSA, 1996a). In terms of local government affairs, the Constitution recognizes it as a distinctive sphere of government and mandates local governments to 'give priority to the basic needs of the community, and to promote the social and economic development of the community; and participate in national and provincial development programs' (RSA, 1996a, p82). It further obliges them to 'encourage the involvement of communities and community organizations in the matters of local government' (RSA, 1996a, p81). These principles, which are spelled out in the Constitution (RSA, 1996a) are reinforced by the Local Government White Paper (RSA, 1998a, p37) (see below), which states that, 'government in South Africa is constituted as national, provincial and local spheres of government. These three spheres are distinctive, interdependent and interrelated. Local government is a sphere of government in its own right, and is no longer a function of national or provincial government. It is an integral component of the democratic state'. This elevated status of local government and the associated participatory rights of citizens is a clear reflection on just how far policy and authority is devolving.

The Local Government White Paper of 1998

The Local Government White Paper (RSA, 1998a) marked a key break from past conceptualizations of local government in South Africa. It translated the objectives for participation and development set out in the RDP and the mandates outlined in the constitution into a definable statement on how local government would look and function in the future. The document clearly argues that 'development local government' is a core focus for local government and that these institutions must work together with their local residents to improve economic and social conditions in the areas under their jurisdiction. In addition, local government is required to take on a leadership role, involving citizens and stakeholder groups in the development process, to build social capital and to generate a sense of common purpose in finding local solutions for sustainability. Local municipalities thus have a crucial role to play as policy-makers and as institutions of local democracy, and they are urged to become more strategic, visionary and ultimately influential in the way they operate. Through undertaking 'developmental local government', it is expected that four key outcomes will be achieved, namely:

- provision of household infrastructure and services, with priority given to the delivery and subsidization of at least a basic level of services to those who currently have little or no access to services;
- creation of livable, integrated cities, towns and rural areas in which the spatial legacy of apartheid separation is addressed;
- achieving local economic development in which local government can play an important role in job creation and in boosting the local economy through the provision of business-friendly services, local procurement, investment promotion, support for small businesses and growth sectors;
- community empowerment and redistribution (RSA, 1998a).

The parallel White Paper on Municipal Service Partnerships details how municipalities can enter into partnership arrangements with the private, public, community and NGO sectors to improve service delivery in a specific area (RSA, 2000b). Since 1998, the principles contained in the Local Government White Paper and the Constitution have been translated into a series of Local Government Acts, which give greater meaning to these principles and devolve the legal powers to enable these concepts to be implemented. These Acts and two key Acts which preceded 1998 have promoted the dual focus on participation and development as outlined in the next sub-section.

Local Government legislation

Although the most recent legal provisions pertaining to the developmental role of local government and participatory governance have been based on the 1998 Local Government White Paper (RSA, 1998a), pre-1998 Acts have also helped to lay a foundation for this new role.

The Local Government Transition Act (RSA, 1996b). This Act provides a basis for development and assigns various powers and duties to local governments relating to service provision. It requires councils to promote integrated economic development, the equitable distribution of municipal resources and delivery of services. Councils are also required to formulate and implement an 'Integrated Development Plan' incorporating land use, transport and infrastructure planning and the promotion of integrated economic development.

The Development Facilitation Act (RSA, 1995). As with the preceding Act, this provides a further basis for development and has introduced measures to facilitate and accelerate the implementation of reconstruction and development programs and projects in relation to land, laying down general principles governing land development throughout the country. Local governments are empowered to develop what are known as 'Land Development Objectives'. These are for the sub-division and development of land in urban and rural areas to promote the accelerated provision and development of land for residential, small-scale farming, economic uses or other needs and to improve security of tenure. This Act was deemed necessary in the light of the complex apartheid geography of the country and the need to redress imbalances and accelerate development through the efficient utilization of land.

The Local Government Municipal Demarcation Act (RSA, 1998b). This Act is concerned with determining new municipal boundaries throughout South Africa, a process undertaken between 1998 and 2000. This Act sought to eliminate small and ineffective local councils through combining neighboring or near local authority areas under a single jurisdiction, and also assigned rural areas surrounding urban centers to the control of the latter. This was undertaken to ensure economic efficiency, such that within municipal boundaries the municipality would be capable of fulfilling its constitutional obligations including the promotion of social and economic development, integrated development and effective local governance.

The Local Government Municipal Structures Act (RSA, 1998c). This Act, along with the Systems Act (see below) extends and develops the provisions of the Local Government Transition Act of 1996 (RSA, 1996b). The Act provides for differing categories of municipality, to operate within the newly demarcated areas and assigns them specific powers and duties. The powers and duties are based on the Constitution (RSA, 1996a) and are generally of a service type nature, but include the following development-type foci: tourism, planning, public works, infrastructure development and markets. In undertaking such duties municipalities are expected to promote economic and social development in the areas under their jurisdiction. The Act also allows for the participation of traditional leaders within local government administration in the areas in which they reside.

The Local Government Systems Act (RSA, 2000a). This is the last key piece of legislation and the one which has the most direct influence over the principle of popular participation in local governance and local-level development. The Local Government Municipal Systems Act (RSA, 2000a, p2) provides for 'the core principles, mechanisms and processes that are necessary to enable municipalities to move progressively towards the social and economic uplifting of communities, and ensure universal access to essential services that are affordable to all'. The Act goes on to detail the government's commitment to the encouragement of participation. It argues that:

> A fundamental aspect of the new local government system is the active engagement of communities in the affairs of municipalities of which they are an integral part, and in

> particular in planning, service delivery and performance management ... there is a need to create a more harmonious relationship between municipal councils, municipal administrations and the local communities through the acknowledgement of reciprocal rights and duties (RSA, 2000a, pp2-3).

Municipalities are specifically required to involve communities in the affairs of the municipality, to provide services in a financially sustainable manner and to promote development (RSA, 2000a). Public participation is a key element of the Systems Act, and municipalities are obliged to establish mechanisms for public participation and participatory governance. The Act makes it a municipal duty to 'encourage the involvement of the local community ... and to consult the community with regards to issues and options for service delivery' (RSA, 2000a, p20). Simultaneously, it is now a community's right 'to contribute to the decision-making process of the municipality ... (and) to demand that proceedings of the municipal council and those of its committees must ... be open to the public' (subject to certain restrictions) (RSA, 2000a, p20). Municipal administrations are obligated to establish a working relationship with the local community and to provide them with accurate information.

Chapter Four of the Act deals specifically with community participation and according to section 16(1) obligates municipalities to develop a system of 'participatory governance' (RSA, 2000a, p30) and 'to create conditions for the local community to participate in the affairs of the municipality' including reviews of development plans, performance management, budget, strategic decisions and, significantly, to build community capacity to enable them to participate in municipal affairs'. A municipality has to consider public petitions and complaints and to allow for public comment, public meetings, consultative sessions and report-back sessions and to make allowances for people who cannot read or write, people with disabilities and disadvantaged groups. In addition, all meetings of the municipal council and their committees have to be open to the public and the media except under special circumstances.

The Act spells out the same powers and duties as detailed in the Structures Act and obligates municipalities to undertake developmentally orientated planning (RSA, 2000a), once again requiring municipalities to develop 'Integrated Development Plans'. These plans should involve widespread consultation with communities and other stakeholders and should link and co-ordinate all municipal development plans, municipal resources, capacity and budgets and be compatible with national and provincial planning requirements (RSA, 2000a). In terms of service provision, municipalities are required to prioritize the basic needs of the community and to ensure that all residents have access to a minimum level of basic services. The Act clearly provides the mandate for participatory governance in local government affairs and development matters.

Critique

Enshrining the principles of participatory governance is clearly one of the most noteworthy achievements of post-apartheid administration and democracy at the local government level in South Africa. This step is, however, very recent and applying the principles in practice will not happen overnight, particularly the notion of citizen participation in a society so familiar with top-down imposition. In this section of the chapter various operational constraints impacting on the principles of participatory governance are discussed.

Whilst upholding the principles of participatory governance is a noteworthy triumph of democracy, the outcomes of such a process cannot be predicted with certainty. Though anecdotal, it is the author's own experience from having participated in a range of local government forums and participatory planning sessions in the late 1990s and early part of the current decade that only a tiny fraction of a local community actually chooses to exercise its rights to participate. Meetings are often characterized by the attendance of a handful of individuals and it would be difficult to say whether, after consultation, their input actually has a fundamental impact on municipal decision-making and implementation. An additional concern includes the fact that participatory governance seems to be interpreted on the ground as consultation about plans, not citizen involvement in project implementation and review. Further, there are no legislated forums or channels for direct participation and involvement in such projects, rather an 'open' system of permitted participation is upheld. The absence of defined mechanisms or means of exercising and enforcing popular decisions is a very real concern. This clearly presents the possibility of reducing civil-society to the role of only being a initial 'rubber-stamp' without any real recourse to principles which might insist on continued participation at all stages of project implementation. In addition, it would be very easy for the authorities to fulfill their duties through paying lip-service to their obligations and engage only in nominal consultation (Lyons and Smuts, 1999). Of equal concern is the fact that even though, as the 1998 Local Government White Paper (RSA, 1998a) states, partnership formation is an option to pursue, it would be difficult to argue that, with the exception of the White Paper on Municipal Service Partnerships (RSA, 2000b), any real effective mechanisms to permit this to be pursued are provided. The Systems Act (RSA, 2000a) encourages participation by civil-society, but, at a broader level, it does not provide any clear mechanism to involve NGOs or the private sector in a defined role. The potential of these sectors to contribute meaningfully to local development is, therefore, not excluded, but is certainly not directly encouraged. Decision-making and implementation power is left very firmly in the hands of local government. Such a development track appears to differ for that pursued in the developed nations and has the potential to perpetuate conventional 'authority' and 'non-authority' divisions at the local level. Naturally, this raises the question of whether this is a viable situation? Whilst the argument could be made that since a local government consists of the elected representatives of the people, such exclusive powers might be appropriate. However, as evidence detailed below suggests, if fundamental capacity and resource constraints exist at the local government level, effecting change through a local government mandate might be severely limiting in scope and potential.

Key considerations which impact not just on the principle of participatory governance at the local government level, but also the ability of local government to meaningful involve local citizens include the fact that a significant percentage of local authorities (an estimated 50 per cent) are bankrupt or in serious financial straits (Nel, 1998), and most small towns simply do not have the available trained personnel to initiate and oversee all the tasks which they are assigned (Ferreira, 1997). In most cases local governments, prior to 1994, consisted of only a handful of officials charged to administer services in the 'Whites' only section of towns, while the majority of people, the Black community, fell under the direct jurisdiction of central government. In the mid-1990s the same officials, with the same limited resources had to take on the responsibility for administering a vastly increased population and physical area with no meaningful extension in their resources, finances and staff. Under such a situation officials are struggling to keep towns functioning and to maintain services and there is seldom any capacity to take on new responsibilities. In addition, by 2001, it was estimated that 62 per cent of municipalities required 'urgent support' given their poverty and inability to render even adequate service provision (Cull, 2001, p11). This obviously falls very far short of the commitment to the more ambitious ideal of 'developmental local government'. As a result, and as on the ground evidence shows, NGO or semi-private local development initiatives might well be a viable development option in many centers (Nel, 1999). The policy shift which has taken place, however, has and could lead to conflicts between legally empowered, but resource-poor, local authorities and other agencies already active in development (Nel and McQuaid, 2000).

Vastly greater levels of funding will be needed if local government-led development, and meaningful participatory governance is to become a more widespread phenomenon and to have a meaningful impact, especially in the light of the financial constraints which local governments currently face. At present, many local authorities simply are not generating enough funds from rates and services charges to cover their costs, let alone embark on new activities. As Emdon (1997, p24) points out 'resources are scarce, and there are limitations'. According to Mukhopadhyay (2000) 'it is ridiculous to tell cash-strapped local governments to manage their own affairs and then expect them to solve the country's problems ... (it) is not the responsibility only of local government ... the (central) government cannot abrogate its responsibilities'. Further, it 'appears that a lot of responsibility has been placed on local councils without an initial program of nation building, including support for local authorities in terms of building capacity' (Mosiane, 2000, p19). Whilst financial transfers are being made to local government, they tend to be more of a crisis response than allowing for forward planning or for significant development, leading Ngcobo (2000) to warn of the dangers of limited delivery being effected.

Based on the preceding discussion it is difficult to avoid the conclusion that there is an evident disparity between rhetoric and actual practice. The principles of participation are sound, mature and well conceptualized. Implementing them, however, is impacted on by the weak status of many local governments at present and the failure of current laws to mandate any tangible form of direct participation in development and planning beyond the level of broad, community-based consultation.

Conclusion

Concluding an analysis of the state of participatory governance in South Africa leaves one in a somewhat ambivalent frame of mind. The principles of direct involvement by the country's citizens in a process which entrenches democracy and popular participation in decision-making are clearly a welcome step forward, particularly after decades of apartheid discrimination. Furthermore, the elevation of local government to the level of being regarded as a distinct sphere of government, having high degrees of autonomy, is also far-sighted and innovative. In parallel, efforts at empowerment, nation building and the promotion of equity and unity, despite historical diversity are all clearly important. However, the failure to create specific mechanisms for participation, together with very real capacity and financial constraints at local government level, will impact negatively on the pace and prospects for change.

Despite these words, and the very real concerns which they allude to, one must not lose sight of the broader picture and just what would have happened in South Africa had this fundamental process of reform not been initiated. This situation is not perfect, but nonetheless is clearly an improvement on past conditions. 'The rainbow nation', coined by Archbishop Desmond Tutu, is a 'phrase (which) captures the extraordinary diversity of races, tribes, creeds, languages, landscapes and living standards - that characterize modern South Africa. But, it is also a euphemism: the after-effects of South Africa's divisive and violent past are still with us today. Notwithstanding the near-miraculous peace-making of the past decade, the country is no model of racial harmony. Yet its various tribes have learned, over years, to live with one another and in some cases to celebrate their differences' (Editors Inc, 2000, p24). South Africa has a very diverse cultural heritage and a troubled past. Enforcing necessary change and equality is agreed to and legally sanctioned. Hopefully, through time, participatory governance and development will be further enhanced to the mutual benefit of the nation and its peoples.

Despite the negative issues which have been raised, the policy shift which is taking place in South Africa is clearly of great significance and, if managed correctly, could lay a valuable basis for addressing the country's very serious development challenges. In sum, 'despite the concerns which have been raised, it is interesting to note that, in contrast to most developing countries, a search is being made for innovative strategies which break with the traditional mould of state-centered planning and which, hopefully, hold some promise for the majority of South Africans' (Nel, 1997, p72).

References

ANC (African National Congress) (1994) *The Reconstruction and Development Program: A policy framework*, Johannesburg: Umanyano Publications.

Binns, T. (1998) 'Geography and development in the 'new' South Africa', *Geography*, 83 (1), pp. 3-14.

Cull, P. (31 October 2001) 'R328 m. to Help out Local Authorities', *Eastern Province Herald*, Port Elizabeth: Times Media Limited, p. 11.

DBSA (Development Bank of Southern Africa) (2000) *South Africa: Inter-Provincial Comparative Report*, Johannesburg: DBSA, Halfway House.

Department of Provincial and Local Government Affairs (DPLG) (2000) *Local Economic Development: Guideline to Institutional Arrangements, Local Economic Development Manual Series*, Volume 1, DPLG, Pretoria.

Economist, The (2001) 'A survey of South Africa', *The Economist*, Vol. 358 (no. 8210), February 24-March 2.

Editors Inc (2000) *SA 2000-01: South Africa at a Glance*, Johannesburg: Editors Inc.

Emdon, E. (1997) *Legal Constraints to Local Economic Development at Local Government Level*, Johannesburg: Friedrich Ebert Stiftung.

Ferreira, N. (1997) personal communication, former Mayor, Stutterheim.

Gooneratne, W. and Mbilinyi, M. (eds) (1992) *Reviving Local Self-Reliance*, Nagoya: United Nations Center for Regional Development.

Haughton, G. (1998) 'Principles and practice of community economic development', *Regional Studies*, Vol. 32, No. 9, pp. 872-877.

Lemon, A. (1987) *Apartheid in Transition*, Aldershot: Gower.

Lester, A., Nel, E.L. and Binns, T. (2000) *South Africa, Past, Present and Future*, Harlow: Longman.

Lyons, M. and Smuts, C. (1999) 'Community agency in the New South Africa: A Comparative Approach', *Urban Studies*, Vol. 36, No. 12, pp. 2151-2166.

Mosiane, N.B. (2000) 'The evolving local economic development process in Mafikeng: a contested terrain between political and profit interests', *South African Geographical Journal*, Vol. 82, No. 1, pp. 13-20.

Mukhopadhyay, S. (2000) *Democratic decentralization and LED: India and South Africa*, seminar presented at Rhodes University, Grahamstown.

Nel, E.L. (1997) 'Evolving Local Economic Development Policy in South Africa', *Regional Studies*, Vol. 31, No. 1, pp. 67-72.

Nel, E.L. (1998) 'Local Economic Development', in Atkinson, D. and Reitz, M. (eds), *From a Tier to a Sphere: Local Government in the New Constitutional Order in South Africa*, Johannesburg: Heinemann, pp. 152-170.

Nel, E.L. (1999) *Regional and Local Economic Development in South Africa: The Experience of the Eastern Cape*, Aldershot: Ashgate.

Nel, E.L. (2000) *Post-apartheid 'Development': Praxis and Policy in South Africa*, paper presented at the Workshop on: South Africa: Past, Present and Future, St. Mary's College, London.

Nel, E.L., and Binns, T. (2001) 'Initiating 'developmental local government' in South Africa: evolving local economic development policy', *Regional Studies*, Vol. 35, No. 4, pp. 355-362.

Nel, E.L. and McQuaid, R. (2000) *Critical Reflections on the Evolution of Local Economic Development in Stutterheim,* paper presented at the Local Economic Development Conference, University of Sussex, Brighton.

Ngcobo, R. (2000) *Regeneration in South Africa*, conference paper delivered at the LED Conference, University of Sussex, Brighton.

Pycroft, C. (1998) 'Integrated development planning or strategic paralysis? Municipal development during the local government transition and beyond', *Development Southern Africa*, Vol. 15, No. 2, pp. 151-163.

Raco, M. (2000) 'Assessing community participation in local economic development - lessons for the new urban policy', *Political Geography,* 19, pp. 573-599.

Republic of South Africa (RSA) (1995) *Development Facilitation Act*, Act No. 67 of 1995.

RSA (1996a) *The Constitution of the Republic of South Africa*, Act No. 108 of 1996.

RSA (1996b) *Local Government Transition Act Second Amendment Act*, Act No. 97 of 1996.

RSA (1998a) *The White Paper of Local Government*, Department of Constitutional Development, Pretoria.

RSA (1998b) *Local Government Municipal Demarcation Act*, Act No. 27 of 1998.

RSA (1998c) *Local Government Municipal Structures Act*, Act No. 117 of 1998.

RSA (2000a) *Local Government Municipal Systems Act*, Act No. 32 of 2000.

RSA (2000b) *White Paper on Municipal Service Partnerships*. Notice 1689 of 2000.

RSA (2001) *The role of a Coat of Arms,* (http://www.polity.org.za/govdocs/misc/cofarms 00.html) 21 October 2001.

Solomon, D. (1990) 'The financing of regional policy: the Regional Services Councils', *South African Journal of Economics*, Vol. 58, No. 2, pp. 257-268.

Stats S.A. (1999) *The Census in Brief*, Pretoria: Statistics South Africa.

United Nations Center for Human Settlements (Habitat) (UNCHS) (1998) *Towards a World Charter of Local Self-Government*, Nairobi: UNCHS.UNCHS, 31 March 2001, Sub-Saharan Africa Supports World Charter of Local Self Government, *UNCHS Press Release,* CHS/00/09, Nairobi: Habitat.

Valler, D., Wood, A. and North, P. (2000) 'Local governance and local business interests: a critical review', *Progress in Human Geography,* Vol. 24, No. 3, pp. 409-428.

Waldmeir, P. (1997) *Anatomy of a Miracle*, London: Viking.

Chapter 3

Local Government, Local Development and Citizen Participation: Lessons from Ireland

Jeanne Meldon, Michael Kenny and Jim Walsh

Introduction

The purpose of this chapter is to provide a guide to citizen participation in local government and governance in Ireland as an aid to fostering debate on best practice in local development. The remit of the local government system in Ireland is to deliver a range of services to the community that it serves, through a democratic system of locally elected councilors. However, it is increasingly recognized that development is a multi-dimensional process incorporating economic, social and environmental objectives, which can only be successfully delivered through the contribution of participatory governance embracing cross-sectoral partnerships.

The issue of participation, or the lack of it, has found its way on to national and international agendas. In many parts of the world, interest in research and practice related to participatory development is increasing. It is now accepted that citizen involvement in local development is the key to equality, inclusiveness and sustainability. It is impossible to establish a universal definition of participation. An understanding of the concept is often assumed; in practice, development actions are often based on differing perceptions of participation and different perceptions of the level and quality of participation being sought, partly because of the lack of experience of effective participation practice. The literature gives a series of definitions of participation ranging from 'token involvement of people', to, 'autonomous decision-making by popular organizations at local level'. Platt (1996) refers to three types of participation by local communities and individuals. These are:

- physical participation - being present, using one's skills and efforts;
- mental participation - conceptualizing the activity, decision-making, organization and management;
- emotional participation - assuming responsibility, power and authority.

Participation can be top-down or bottom-up, uniform or diverse, simple or complex, static or dynamic, controllable or uncontrollable, predictable or unpredictable. By introducing professionals, controls, bureaucracy and systems, participation can lose its spontaneity, its flexibility and its usefulness. An accurate definition needs to accommodate the complexity inherent in participation and the power relationships that enable or hinder participation. Reversing power is the key to participation. 'Putting people first in development projects is not just about organizing people but it means empowering them to be social actors rather than passive subjects and to take control over the activities that affect their lives' (Cernea 1985).

Participation is generally considered a core value in community development. While community development has for a long time been recognized as a beneficial process, the importance of participation within community development has been inadequately stressed. This is partly due to the lack of a clear interpretation of development, and, therefore, of the key constituents of effective development. The need for a new paradigm to address this deficiency was highlighted by the UNDP in its *Annual Human Development Report* (1994). The UNDP's paradigm:

> Puts people at the center of development; regards economic growth as a means and not an end; protects the life opportunities of future generations as well as the present generations.

The associational sector has been referred to as the innovator of participatory research and development. Third World development approaches have long recognized the centrality of participation not only as a development strategy but also as a development objective. However, in our 'developed' societies we have not accepted the importance of participation to the same extent until now. According to one expert in the field of community development, 'There is a fundamental redefinition occurring in Irish society ... a new vision for the future ... happening at the fringes of Irish society rather than at its center ... at the bottom rather than at the top ... more likely to be found in community groups rather than in universities; amongst women than amongst men and amongst voluntary groups rather than in state bodies' (Collins quoted in Reynolds and Healy, 1993, p 103). This shift is being pushed by the empowerment of citizens through local development initiatives, the general increase in awareness and the growth in community based approaches to development which reflect the failure of conventional services to solve problems. The work undertaken in local development in Ireland has provided a laboratory to experiment with varying processes of participation. However, little, if any, critically reflective energy has up to now been invested in identifying the models of good practice. As Ireland moves rapidly into post modernism with the introduction of new, integrated and inclusive structures for local development planning and management, we may see the dawn of a new, highly participative society.

Accordingly, this chapter will present an overview of citizen participation in local government and governance in Ireland using three case studies of good practice. First, local government structures together with mechanisms for participation are outlined in the following section. The emergence of local development structures outside of the local government system are then summarized in order to illustrate the scope for direct citizen engagement in the development process. The response by central government to this challenge is examined through a reform of local government designed to enhance participatory democracy. The chapter concludes by highlighting some key lessons learned from the Irish experience thus far.

Local government structures

The local government system in Ireland includes the local authorities and the regional authorities. The elected local authorities are the county councils (29), the five county borough corporations (representing the larger urban centers), five borough corporations, urban district councils (49) and boards of town commissioners (30). The members of these authorities are elected by a system of proportional representation, with elections taking place every five years. Each constituency is sub-divided into a number of local areas from within which candidates are elected to sit on the council. The membership of county councils and county boroughs ranges from 15 to 52. Borough councils have 12 members and urban district councils and boards of town commissioners have nine members. The county councils and county boroughs are the principal agents of public administration with a lesser range of functions coming within the ambit of the other bodies. Town commissioners have a more limited range of functions than the other authorities. The principal services provided by the local authorities comprise:

- Housing and building
- Road transportation and safety
- Water supply and sewerage
- Development incentives and controls including land use planning
- Environmental protection
- Recreation and amenities
- Agriculture, education, health and welfare
- Miscellaneous services

Compared to other European states, the Irish local government system is relatively weak with a more limited range of functions and powers. Local authorities have no role in policing, public transport or personal social services. Powers in respect of education, health and agriculture have been severely constrained and the only social function is in respect of housing.

The functions of local authorities are divided into executive and reserved functions. The reserved functions are performed directly by the elected members and comprise major matters of policy and principle. Such functions include the adoption of annual estimates and borrowing of money; the making of development plans; making, amending or revoking bye-laws; bringing enactments into force and nominating persons to act on other public bodies. Any function that is not a reserved function is the responsibility of the executive. The policy role of the elected members has not been fully realized, partly due to the lack of adequate support systems.

The citizen and local government

Structures for participation in local government are in a state of flux in Ireland with the recent establishment of Strategic Policy Committees and County Development Boards. These structures are examined in more detail below. Historically, apart from representation through the democratic process, citizens have participated directly in local government through a limited number of committees and through the planning process.

Planning is a significant function of the local authority. Day-to-day decisions on individual planning applications are an executive function while the adoption of Development Plans is a reserved function for elected representatives. Historically under the legislation and structures a draft Development Plan is prepared by a council's staff and is put before the elected members for consideration. The draft is adopted and put on public display for three months, prior to its final ratification[1] by the council. It was customary for the opportunity for public participation to be largely limited to reaction to a draft Plan. Such involvement may be termed passive consultation. However, the terms of the recent 2000 Planning Act enhance the scope and range of participation for citizens in the planning process. There is now a statutory obligation on the planning authority to advertise its intention to prepare a draft plan and to invite comments and submissions. Individuals, community groups or organizations also have a right to make submissions on planning applications. Under the 2000 Planning Act, the manager / chief executive is required to take into account any objections lodged on planning applications, whereas previously they may or may not have been taken into account by the planning authority.

Despite the limitations inherent in the planning system, there are some examples where citizen participation has extended beyond the scope of the regulatory framework thus pointing to opportunities for developing models of good practice for more active involvement. Case Study One is such an example, but many others could be cited, including Planning for Real approaches where the local authority, through the use of physical models, has invited people in a community to explore a range of options for their area.

CASE STUDY ONE: CITIZEN PARTICIPATION IN LOCAL AREA PLANNING IN COUNTY CLARE

Overview

Ballinruan/Crusheen are two townlands five miles from the county town (Ennis) in the Mid-Western coastal county of Clare, in the West of Ireland. These rural based townlands have just fewer than 400 houses. The townlands are characterized by high population growth on the one hand and pockets of severe rural disadvantage on the other. There are a number of abandoned houses and a significant number of houses with single elderly occupants. The local economy is strongly supported by an active working population who travel out of the area for work in Ennis or further afield. The local development council became very concerned that the rural setting of these townlands would be radically altered by the spillover of housing from Ennis. They feared, from the experience of many other rural communities on the outskirts of fast developing urban centers that permissions for housing development would be granted without consultation.

Because of the concerns of the local group, these townlands were selected as pilot areas for a Clare LEADER[2] initiative called the Local Development Planning Program. The objectives of the program were:

- To engage maximum participation of the community in the drawing up of the plan
- To provide relevant training to build the capacity of the community in the formulation and implementation of the plan

This program is an innovative and comprehensive approach to the assessment of needs by communities in rural areas. In 1997 the parish of Crusheen/Ballinruan together with Rural Resource Development (RRD) and Clare County Council came together to formulate a local development plan for the two townlands. The participatory planning process that emerged was very innovative in the context of local development and is a model of good practice on positive citizen participation.

Process

Following initial meetings the partners in this process proposed to use the 'house meeting'[3] model as a way of jointly assessing needs. The committee planned the process, prepared work sheets, called a briefing evening for house meeting hosts and leaders, oversaw the carrying out of the process and then through a workshop assembled the feedback into a booklet. Sixteen house meetings were held with almost one hundred and fifty people attending. A wide range of occupations and ages were represented. The feedback from the house meetings was very positive mainly because they were friendly and informal. There were high levels of participation. People met as neighbors and friends to discuss the issues that concerned them and expressed a wish to do so again. At a second round of house meetings, nominations were taken for a Steering Committee, made up of stakeholders in the local planning process, to oversee the planning process and to take responsibility for the many aspects of it. This Group, including the volunteer local development grouping, met to agree collaborative actions for more effective partnership in meeting needs and improved quality of life in those rural communities. Thirty-four people became actively involved in the process. This planning forum met regularly with active participation to develop a local development plan to meet local needs.

In the following months, a vision statement and objectives were prepared and the professional planner was engaged to assist with the process. This involved formulating and devising a questionnaire to gather information from the people of the community which, following analysis, would identify the options for project and program development in the local area. The next stage of the process was to prioritize options from the analysis and to make a draft plan available to the general community for its consideration. During the process, the community was consistently kept informed of progress. This was an important factor in ensuring an almost 87 per cent response from all households through a completed questionnaire.

Outcomes

The house meeting consultative process and the planning forum is deemed to be a very good model of local participation and partnership in planning local development. The planning forum, which continues to meet, is a very open process of two-way information exchange of information from Clare County Council and the local areas. A high level of trust has developed with recognition of the openness and flexibility of Clare County Council planning officials. As the process is ongoing, the outcomes in terms of more acceptable decisions, more effective service delivery and more sustainable local development structures have yet to be proven. An integral component of the local development planning process was the introduction of the capacity building program *Training for Transformation.* It has made the difference in building the capacity of the community by recognizing its own knowledge and experience and by empowering participants to take action on matters relating to community life.

The process is being replicated in the adjoining parish of Feakle and a similar process is underway. The experience of the pilot project in Crusheen/Ballinruan was of huge benefit to all of the partners involved, in this latest initiative, and while some changes have been made to the original process, the overall model and objectives have been adhered to. The entire process has taken more than two years to complete, with huge commitment from both communities. In short, the local development planning model brings the term “partnership” into a new dimension, which allows for participation, consultation and transformation.

Citizen participation and local governance

The emergence of area based local development structures outside of the local authority system has taken place in the context of a growing realization globally, nationally and locally that to be sustainable, development should bring about not only an improvement in social and physical conditions, but it must also contribute to an improvement in the capacity of people and communities to control and sustain those conditions.

The limitations of both central and local government systems together with the perceived failure of the statutory agencies to address persistent problems of urban unemployment and disadvantage and of rural deprivation, led to the first pilot area-based local development initiatives in the late 1980s. The late 1980s also saw the development of the social partnership model at national level and its success was one of the reasons for the extension of the partnership model to the local level in the early 1990s. The brief of the partnership companies was to work with the long-term unemployed and socially excluded, *i.e.* those most marginalized from economic and social life.

EU funding mechanisms under the Structural Funds provided the opportunity for further expansion of such initiatives during the period of the first Community Support Framework (CSF) for Ireland, 1989-1993. Following the initial operation of 12 pilot partnership companies, sixteen area-based partnerships were established to tackle long-term unemployment with the support of a global grant administered through ADM (Area Development Management). County Enterprise Boards (CEB) were established on a pilot basis to promote and support micro-enterprises. In addition under the EU LEADER 1 Community Initiative, 16 LEADER groups were established in selected areas of the country to promote an area-based approach to rural development.

Under the terms of the Operational Program for Local Urban and Rural Development 1994-1999, further funding was made available to deliver integrated programs on (i) local enterprise delivered through the mechanism of the CEB on a countrywide basis, (ii) integrated development of designated disadvantaged and other areas to be delivered largely through 38 Area Partnerships (see Case Study Two below), and (iii) urban and village renewal. These measures were to complement the funding made available through the LEADER II Community initiative (1994-1999), which provided funding for rural development groups, also on a countrywide basis.

The objective of the Operational Program was to bring about social and economic development at local level, to involve and to enable local communities to be involved in that development in a formal way and to achieve physical improvements to the environment (OPLURD 1994-1999). Significantly the Operational Program (1994-1999) acknowledged the particular importance of direct participation in the planning and implementation of local initiatives. Participation is achieved through the partnership structures, which are a feature of all of these boards (CEB, Partnership Companies and also LEADER groups). The Boards include representatives from a multiplicity of sectors and social partner organizations including state agencies, local authorities, employers' organizations, trade unions, farming organizations as well as voluntary and community based organizations. The availability of resources is important for small-scale local projects that both support local entrepreneurs and fund activities which contribute to an improvement in the quality of life, and enhance the capability of communities to participate in the process.

The program recognized the role which local initiatives can play as a catalyst for local economic, social and environmental development and the importance of locally based measures to complement the national approach. For example, the role of the CEB is to develop indigenous potential and stimulate economic activity at local level, through the provision of financial assistance and technical support for the development of micro enterprises undertaken by individuals, firms and community groups and support the creation of an enterprise partnership at local level between the social partners, financial institutions and local communities that had not been achieved at local government or statutory levels.

CASE STUDY TWO:
SOUTHSIDE PARTNERSHIP – CITIZEN PARTICIPATION

Southside Partnership is one of 38 area-based partnerships throughout Ireland, and one of 11 in Dublin, supported by Area Development Management and the Operational Program for Local Urban and Rural Development (part of the Irish Government's Community Support Framework 1994-1999) under the Social Inclusion measure. The Board consists of 22 Directors, four representing the social partners, nine representing the community sector, six representing state agencies and three representing the local authorities. The company consists of a number of committees and working groups including the Operating Group (a sub-committee of the Board), Local Employment Services Management Committee, Expanding Economic Opportunities Network and the Education Working Group.

Southside Partnership's Equality Statement states that it "is committed to working towards the elimination of poverty and social exclusion through the achievement of equality of opportunity, participation and outcomes for all people who live in the Southside Partnership's designated areas. Southside Partnership's work is informed by a human rights philosophy". The goal of Southside Partnership is to tackle poverty and exclusion in pocket communities in the Dun Laoghaire Rathdown and South Dublin County Council areas. Its target groups and communities experience extreme poverty, often characterized by economic deprivation which carries on from one generation to the next, high levels of welfare dependency, educational under-achievement, poor quality environment and housing, inadequate service provision and problems of vandalism, drug misuse and crime. While the various target groups such as the long-term unemployed, Travelers, lone parents and people with disabilities have differing characteristics and problems, they share the common experience of being marginalized from society and having little control over the decisions and policies that shape their lives.

The work of the Partnership, therefore, is rooted in a community development perspective that seeks to empower marginalized groups to take control of the decisions affecting their lives. Citizen participation is central to all aspects of its work and is achieved in the following ways:

Consultation: An extensive consultation process was undertaken for the Partnership's six-year strategic plan 'United Vision 2000-2006', involving over 1000 people in developing a vision for the future of their communities. Target groups were involved in the design and hosting of the consultation and innovative and creative methods were used to maximize participation. Draft plans were devised on the basis of people's input and then fed back to the community for final agreement, thus ensuring consensus on the strategies adopted.

Capacity Building: The Partnership works with target individuals, groups and communities on an intensive basis to strengthen their capacity to effect change in their own lives and their communities. Groups are facilitated to develop plans for their areas and work in collaboration with outside agencies to implement them. A strong emphasis is placed on leadership development to train community leaders to represent the interests of disadvantaged groups at policy level. A Local Development Training Institute (LDTI) has been set up by Southside Partnership to facilitate this process. Over 1400 activists have received training at LDTI since 1997.

Partnership structures: The Partnership Board is composed of representatives of social partners, state agencies and community and voluntary interests. This blend of interests and perspectives is replicated throughout all Partnership structures. Over 30 networks and working groups have been established to facilitate all sectors to work together to devise and implement solutions to issues such as early school leaving, enterprise development, employment services, Traveler inclusion and community development support. These structures challenge and facilitate decision-makers to change their practices and policies to ensure that local people have control over the decisions affecting their communities.

For Southside Partnership, citizen participation is not an end in itself, nor is it restricted to programs and activities that are Partnership initiated. It is about challenging the non-inclusive way that decisions have traditionally been made for disadvantaged communities and ensuring that structures are put in place to guarantee effective ongoing participatory planning. It is also about strengthening people's ability to engage as equal partners in the process.

One example of the Partnership's approach to promoting active citizen participation is its Community Development Support Program (CDSP). Thus far, this program has brought together a range of statutory agencies and local development groups to devise a strategic program of support for community development in the region. As a consequence of the CDSP, there has been a proliferation in the number of community projects and their capacity has been strengthened. The involvement of people from all target groups in community initiatives has greatly expanded. Furthermore, the CDSP has facilitated the local community sector to strengthen its collective voice through the establishment of a community platform and other networks and has established an inter-agency forum to facilitate dialogue between state agencies and community representatives. Over the period 2000-2006, this work will be further developed, and a strong emphasis will be placed on the improvement of accommodation and physical living conditions of disadvantaged communities, which emerged as the principal issue of concern in the consultation process.

The reform of local government – enhancing citizen participation

Local government in Ireland has developed largely from a judicial system introduced under a colonial regime and is historically more removed from its community base than many continental European systems of local government. The local government system is inhibited by a lack of resources and an over-dependence on central government decisions made annually as part of the budgetary process, and by a lack of coherence and co-ordination in the delivery of services. Through lack of resources and an inability to respond to new pressures that transcend their traditional functions, local authorities have to some extent been by-passed by the growth of the partnership companies (CEB, LEADER Groups and Partnerships) many of which excluded local government representation, thus creating tensions between two largely separate local development systems.

Traditionally citizen participation in local government has been through the electoral system, with councillors representing each local electoral area. The elected members receive their mandate to represent citizens through the democratic process of local elections held every five years. Despite the democratic process inherent in the system, local communities have felt increasingly alienated from local government. This is partly because of frustration with the limited range of activities which fall within the remit of the local authority and perceived inefficiency, partly because of the clientelism associated with elected members who are perceived as representing the views of only some sections of the community linked to the perceived corruption of local government, and partly because of the perceived failure of local authorities to be pro-active in responding to new needs and demands. At the same time local councillors have felt frustrated by the range and scope of activity, and the funding opportunities made available to local development groups, which seemed in many cases to be taking over the functions that should accrue to democratically elected representatives.

In response to these factors and also because of the recognition that the introduction of social partnership in Ireland at national level has provided the basis for social and economic progress, proposals for enhanced participative democracy at local level were set out in a 1997 program for *Better Local Government*. The Devolution Commission set up by Government in 1995 to address the problem of uncoordinated and exclusive systems established the basic principle behind the reform, i.e. the existing local development systems should be brought together. The *Better Local Government* program set out to ensure that:

- local communities and their representatives have a real say in the delivery of the full range of public services locally;
- new forms of participation by local communities in the decision-making processes of local councils are facilitated;
- the role of local councillors in setting policy and giving leadership to socio-economic development together with the social partners is enhanced.

Measures were proposed which would recognize the legitimacy of local government as a democratic institution, enhance the electoral mandate within local government and broaden involvement in local government. The reform measures have included the establishment of Strategic Policy Committees (SPC) within the local authority system, which bring together elected members, community representatives and sectoral interests. The establishment of the SPC give councillors a more meaningful role in policy review and development, offer an opportunity for the deeper involvement of councillors in the corporate governance of the local authority and provide an enhanced opportunity for citizen involvement. Elected members chair the SPC and councillors are in the majority on each committee. However, at least one-third of the members of each committee are drawn from bodies within the community that are relevant to the work of that committee. SPC can provide advice/reports. directly to the Council on matters within their range of responsibilities or on related issues, either on their own initiative or as decided by the Council. Essentially SPC represent a restructured committee system within each local authority.

The program was further revised following consultation and negotiation on the introduction of new structures, and the recommendations of an interdepartmental Task Force that had been established to put forward a model for the integration of local government and local development systems. The Task Force[4] put forward a set of key principles (Table 3.1) to underpin a new model of local governance:

Table 3.1 Local Governance Principles in Ireland

Partnership/participation	A new model will have to be constructed on partnership lines with the meaningful involvement of all sectors;
Social inclusion	The focus on social inclusion of many of the local development structures must be retained;
Community development	The new model should provide an outlet for the involvement of local communities in decision-making;
Democratic legitimacy	Moves towards an integrated framework; needs to recognize the democratic legitimacy of local government while building on the opportunity for more effective participation based on the partnership model;
Process	A new model should be performance driven, should facilitate initiatives, and should recognize the importance of the process of participation for marginalized groups;
Flexibility	The new model should operate with flexibility and an ability to respond to changed circumstances; established public bodies need to be capable of adopting innovative approaches to local and community issues;
Simplicity	The functions and roles of bodies at local level should be clear-cut and as simple as is feasible;
Thematic & area-based approaches	The two should be combined for maximum affect;
Voluntary effort	Existing voluntary effort should be retained and harnessed;
Local Government	Must become participative as well as representative; linkages with other programs including the delivery of EU programs.

These recommendations led to the establishment of County Development Boards (CDB) with effect from 1st January 2000 as a further part of the integration process between local government and local development at county level. The composition and functioning of the Boards is intended to reflect the set of principles set down by the Task Force. These boards will replace County Strategy Groups, which previously had a limited coordinating role for some local services. The primary functions of the CDB are to draw up a comprehensive strategy for economic, social and cultural development within each county and to oversee its implementation. The strategies, which were due to be finalized by January 2002, comprise an agreed framework of overall objectives and targets into which relevant plans can fit rather than a detailed sector-by-sector plan. Responsibility for the delivery of the latter remains with the relevant partners, including sectoral agencies and partnership companies. In addition to the involvement of the CDB in the preparation of these strategies, *ad hoc* working groups covering a variety of themes appropriate to particular counties, such as social inclusion, adult learning, rural transport, tourism, spatial planning and others will be set up to facilitate consultation and participation. These groups will include community representatives. It is envisaged that CDB will have a particular role in ensuring co-ordination of local service delivery. In short, the CDB will develop a vision at local level to encompass various local and sectoral plans, provide the focus for co-operation on a continuing basis at county level in the work of the various agencies, promote co-ordination and, by bringing together the various interests, seek to maximize the effectiveness of spending on those programs and projects at local level included in the National Development Plan.

Directors of Community and Enterprise, appointed within each county in 2000, have drawn up a register of community and voluntary organizations with an interest in joining the Board and its support structures. Voluntary and community groups whose interests include community and rural development, conservation, culture, heritage, Gaeltach,[5] community infrastructure, the environment, recreation, disadvantage, the disabled and others have been invited to register with their County Council in order to join a Community and Voluntary Forum for each electoral area and from which representatives have been nominated to the Boards.

Accordingly, in terms of citizen participation the most important sectors represented on the SPC and CDB are the public representatives and the representatives of the community and voluntary sectors in each county (see Case Study Three below). This process is the first time that the community and voluntary sectors have, as a matter of government policy, been invited as full partners to participate in strategic planning at county level. It is a very significant step in enhancing citizen participation and supporting representative democracy with participative democracy.

CASE STUDY THREE: COMMUNITY REPRESENTATION – MEATH COUNTY COUNCIL STRATEGIC POLICY COMMITTEES & COUNTY DEVELOPMENT BOARD

This case study will detail the process of selecting and electing representatives from the community and voluntary sectors in County Meath to participate in both the SPC and CDB.

Process

The first step in selecting community and voluntary representatives was to consult the community and voluntary sectors in County Meath regarding local government reforms and to inform them about the opportunity for their participation. This task was undertaken by the newly established Office of the Director of Community and Enterprise in County Meath. This is a local authority contract appointment.

A public information meeting targeted at all those groups and individuals active in the community and voluntary sectors in County Meath was called in mid February 2000. At that meeting it was agreed that the following process would be utilized in relation to the establishment of the County Meath Platform for the Community and Voluntary Sectors and the selection of representatives to sit on the Meath County Development Board (CDB) and the four Strategic Policy Committees (SPC) to be established by Meath County Council:

- registration of all community and voluntary sector groups for the purpose of making nominations. A set of criteria to determine those eligible to register was agreed;
- nominations for the positions on the CDB and SPC were invited according to agreed procedures. Nominations were submitted to the Office of the Director of Community and Enterprise by an agreed date;
- at the first meeting of the newly formed County Meath Platform for the Community and Voluntary Sectors nominees were invited to speak for 5 minutes on their reasons for going forward and what they would bring to the role. The meeting also facilitated a roundtable of nominees to discuss issues;
- a second meeting of the County Meath Platform for the Community and Voluntary Sectors was called to ratify the nominations and agree the voting process;
- ballot papers were prepared by Meath County Council officials and were sent to all registered voting groups registered with the County Meath Platform for the Community and Voluntary Sectors and the Office of the Director of Community and Enterprise;
- Meath County Council officials supervised the voting process that was held in conjunction with the next meeting of the County Meath Platform for the Community and Voluntary Sectors. Voting was conducted

according to the proportional representation system and results were announced that night at the meeting;

- elected candidates met as a group with an external facilitator to clarify roles, to determine a collective process and to agree priorities. A process of feedback through the County Meath Platform for the Community and Voluntary Sectors was agreed and a support mechanism for these voluntary elected representatives was proposed.

Outcomes

The process was implemented as agreed and proved both effective and educational. As an outcome, the Platform is established and nine people from the community and voluntary sectors have taken their places on the SPC and CDB. They are working as a team and have the Platform as their forum for consultation within the sectors on an on-going basis. The inaugural meetings of the CDB and the SPC were held and work is ongoing.

The feedback from the information dissemination process, the nomination process, the election process and the team working of the elected representatives was very positive. However, it still remains for the effectiveness of the community and voluntary sectors' representatives on the CDB and SPC to be demonstrated. Equally the effectiveness of the feedback mechanism to the County Meath Platform for the Community and Voluntary Sectors, and the effectiveness of this Platform, have still to be proven.

It must be noted that the above process marks a significant step forward in terms of citizen participation, particularly citizen activists in the community and voluntary sectors, because:

- it is the first time that the community and voluntary sectors have been recognized officially as key partners in local development at county level;
- it is the first formal organization of activists in the community and voluntary sectors at county level;
- it is the first integrated representation of the community and voluntary sectors on the committees of the County Council;
- it is the first time that the local authority has committed to providing resources to support the organization of the community and voluntary sectors at county level.

This process is a strategic recognition of the learning and work achieved through local development organizations in Ireland during the 1990s and the impact of EU funding to strengthen the community and voluntary sectors. It is also the first applied recognition of the role and relationship of participative democracy side by side with representative democracy, and it is the first real challenge to the community and voluntary sectors to participate fully and equally with other stakeholders in the preparation of, and subsequent delivery of, a detailed strategic document for the integrated development of the whole county.

At a broader level a number of issues need to be addressed to ensure that community representatives will be able to participate fully. The level of readiness to participate in such structures varies significantly with some representatives well used to dealing with state agencies in the course of their work, while others have no such experience and may feel over-awed by the process. Gaps exist in the level of knowledge of how the local authority works. In the view of a local government official, the Strategic Policy Committees offer potential for enhanced citizen participation, but it is expecting a lot from voluntary people to contribute to complex issues such as social housing, planning, etc. especially as less and less people are willing to volunteer. The training referred to above will help to overcome some of these barriers. There is also a resource issue for community representatives because of the time input involved for members representing a voluntary committee that would not have either the financial or administrative resources to support that person. Holding meetings in the Council Chamber is an issue for some. This is regarded as councillors' territory and some of the other sectors feel at a disadvantage. The degree of acceptance by other sectors of the involvement of the community sector varies from potential opportunity to intrusion.

The development of new structures to enhance citizen participation in local government and to aid the integration of local development systems has also had to be accompanied by new and innovative measures to improve the delivery of local services. In this vein, a number of local authorities have embarked on a decentralization program to bring those services, which require direct contact between citizens and staff, closer to citizens. For example, in County Donegal, six new council offices are planned with three already built. In the case of County Donegal, the new offices will be staffed by the appropriate personnel actually dealing with the service involved – such as land use planning, motor tax, waste collection, connection to water services, etc. It is envisaged that staff working at local level will be delegated to take decisions in respect of certain functions. Space is being provided in the new local offices to accommodate area based staff from local development agencies and partnerships.

Promoting citizen participation – lessons learned

The evidence above gives some indication of the scope of participatory democracy in Ireland at the present time. We can see from the typology of participation presented in Table 3.2 that citizens may participate, and that agents of change may offer opportunities for citizen participation, for different motives.

Table 3.2 Typology of Participation

Typology	*Characteristics of each type*
Manipulative participation:	Participation is pretence with people's representatives on official boards but who are un-elected and have no power.
Passive participation:	People participate by being told what has been decided and has already happened. It involves unilateral announcements by an administration or project management who do not listen to people's responses. The information offered belongs only to external professionals; example – the traditional County Development Plan process.
Participation by consultation:	People participate by being consulted or by answering questions. External agents define problems and information gathering processes and so control analysis. This process does not concede any share in decision-making and professionals are under no obligation to adopt people's views; example – delivery of local services through One Stop Shops.
Participation for material incentives:	People participate by contributing resources, e.g. labor, in return for material incentives.
Functional participation:	External agents see people's participation as a means of achieving project goals, especially reductions in costs. People may form groups to meet pre-determined objectives. This participation may be interactive and may involve shared decision-making, but tends to arise only after external agents have made major decisions. Local people may only be co-opted to serve external goals; example SPC and CDB, but these are moving towards Interactive Participation.
Interactive participation:	People participate in joint analysis, development of action plans and the formation, or strengthening, of local institutions. Participation is seen as a right, not just as a means of achieving project goals. The process involves inter-disciplinary methodologies that seek multiple perspectives and make use of structured and systematic learning processes. As groups take control over local decisions and determine how local resources are used, so they have a stake in maintaining structures and practices; example, LEADER companies, partnership companies elements of which extend into self-mobilization (see Southside Partnership), partnership in planning (see Ballinraun/Crusheen, County Clare).
Self mobilization:	People participate by taking initiatives, independently of external institutions, to change systems. They develop contacts with external institutions for the resources and technical advice that they need, but retain control over how the resources are used.

(*Source:* adapted from Pretty, 1995)

While much of the experience of citizen participation up to now has fallen far short of the ideal there is an increasing recognition of the need to involve citizens in the process of decision-making. Equally the transition to more participative structures will take time. In order to draw lessons from the experience of participation in Ireland it is necessary to revisit the question of why participation is an essential building block in local development. Progress towards sustainable development requires co-operation and consensus and the participation of all actors in society including the associational sector. Without such active involvement it will not be possible to bring about the kinds of societal change needed to make a real difference. The participation of citizens in decision-making is needed to ensure that social and environmental as well as economic dimensions are included in the process.

Traditionally the local government system in Ireland engaged citizens in only limited, passive participation. It was no coincidence that the local development structures put in place to counteract the inadequacies in the local government system introduced different and innovative opportunities for citizen engagement. The scope for involvement in the processes of local government and governance has been significantly extended through a number of initiatives as outlined above and especially through the partnership bodies. Most recently, the establishment of County/City Development Boards has been an attempt to integrate local government and local development structures and to enhance participatory democracy.

A number of lessons can be drawn from the analysis and case studies in this chapter:

- the transformation from government to governance is a gradual process requiring on-going support;
- objectives for participation leading to a more sustainable model of local development can be achieved in a variety of ways provided that structures are horizontal and interactive, rather than hierarchical and unidirectional;
- institutional structures are important to ensure that all actors are brought into the process; otherwise there is a reliance on *ad hoc* measures and once-off experiments with no lasting impact. The new County Development Boards are an example of an institutional framework which may help to deliver the required outcome, i.e. an integrated strategy for sustainable local development based on principles of inclusion and equality;
- pilot actions can play a significant role, but mechanisms for mainstreaming successful actions are needed. Otherwise there is a risk that the impetus will be lost and that old practices will be resumed;
- meaningful participation requires a commitment of resources to give all participants an equal opportunity and an equal footing as stakeholders;
- openness, dialogue and flexibility are essential to sustain citizen participation;
- meaningful participation requires a change in culture among formal organizations and a change in expectations among citizens;

- capacity building, *i.e.* training for empowerment for those traditionally excluded and training for participation for those who traditionally have held the power is an on-going requirement to move towards interactive participation and self-mobilization;
- evaluation of actions must be built into the process from the beginning. Not all new experiments in participative democracy will work (for example, some models may not be fully inclusive);
- the additional benefits from the partnership model should be clearly reflected in the outcomes, *i.e.* the achievement of new objectives because of the inclusion of all partners in the process;
- the catalyst for change has come partly from within, but has also been driven by the opportunities and requirements of EU programs and instruments;
- bringing together processes that have been devised at local level and with strategic vision from the center offer the best opportunity for movement towards a sustainable model of development.

Conclusions

This chapter demonstrates that there is an increasing level of participation by the associational sector in decision-making structures at local level. Participation as we have seen can take a number of different forms. The next step is to ensure that on-going support is made available to ensure that such participation will contribute to enhancing the foundations for a more sustainable society. Ireland must "hasten slowly" and implement the principles of participation from the start if we are to build sustainable foundations for true citizen participation in all aspects of development.

Most of the structures that have been put in place in Ireland are too recent to be able to fully evaluate their impact. Many other innovative initiatives have been of a temporary pilot nature and have not been subject to full evaluation of either methodology or outcomes. Even where the process is clearly of value, the initiative cannot make a lasting contribution to enhancing participation unless a decision is made to mainstream and to provide on-going funding. Monitoring and evaluation of both local development initiatives and local government structures are required to ensure that the participative process is inclusive and that the outcomes will make a difference.

Acknowledgment

This chapter was compiled as part of the CREADEL Transnational project funded by the European Union INTERREG IIC Program.

References

Area Development Management Ltd. (ADM) (1996) *Local Development Strategies for Disadvantaged Areas: Evaluation of the Global Grant in Ireland (1992-1995)*, Dublin.

Area Development Management Ltd. (ADM) (1999) *Old Problems New Solutions, Building a Better Future for Disadvantaged Groups in Rural Ireland*, Dublin.

Cernea, M. (1985) 'Sociological Knowledge for Development Projects', in *Putting People First: Sociological Variables in Rural Development*, Oxford: Oxford University Press.

Collins, T. (1988) *Community Enterprise: Participation in Local Development*, Unpublished Phd Thesis, St. Patrick's College, Maynooth, Ireland.

Department of Environment (1997) *Better Local Government*, Dublin.

Department of Environment and Local Government (1998) *Report of Interdepartmental Task Force on the integration of Local Government and Local Development Systems*, Dublin

Department of Environment and Local Government (1999) *Preparing the Ground: Guidelines for the Progress from Strategy Groups to County/City Development Boards* Dublin.

Department of Environment and Local Government (2000) *Revised Draft Guidelines for Local Agenda 21* (Working Draft, unpublished)

Gallagher, P. (2000) *The County Development Strategy and the County Development Plan: The role of the new County Development Boards*, Paper delivered to Irish Planning Institute Annual Conference, May 2000, Westport, Co. Mayo.

Government of Ireland (1994) *Operational Program for Local Urban and Rural Development 1994-1999*, Dublin.

NESF (1997) *A Framework for Partnership – Enriching Strategic Consensus through Participation*, Forum Report No. 16, Dublin: NESF.

NESF (1999) *Local Development Issues*, Forum Opinion No. 7, Dublin: NESF.

OECD (1994) *Development Assistance Committee Report on Development Co-Operation*, Paris: OECD.

OECD (2000) *Local Partnerships in Ireland*, Report prepared by Ivan Turok.

OECD/LEED Program (2000) *Draft Report on Local Partnerships*.

Platt, I. (1996) *Review of Participatory Monitoring and Evaluation*, Report prepared for Concern Worldwide, August.

Pretty, J. (1995) 'Participatory Learning for Sustainable Agriculture', *World Development*, Vol. 23, No. 8.

Reynolds, B. and Healy, S. (eds) (1993) *Power, Participation and Exclusion*, Dublin: Conference of Major Religious Superiors (CMRS).

Reynolds, B. and Healy, S. (eds) (2000) *Participation and Democracy, Opportunities and Challenges*, Dublin: Conference of Major Religious Superiors (CMRS).

Shannon Development (1999) *Building Local Development*, Report from a Workshop Series, Shannon.

Thompson, J. (1995) 'Participatory Approaches in Government Bureaucracies: Facilitating a Process of Institutional Change', *World Development*, Vol. 23, No. 9. pp 1521-1547.

United Nations Development Program (UNDP) (1994) *Annual Human Development Report*, Oxford: Oxford University Press.

Notes

1. There are related procedures if the Draft Plan is materially altered as a result of amendments resulting from the public display.
2. LEADER is a Rural Development Initiative of the European Union and is defined as 'Links between areas for the development of the rural economy'.
3. This model builds on the concept of *house stations* in the west of Ireland where traditionally the residents of a townland gather in one house in the townland on a rotational basis for a Mass in the house.
4. Report of the Task Force on the Integration of Local Government and Local Development Systems, August 1998, DOELG, Dublin.
5. Designated Irish speaking areas.

Chapter 4

The National Rural Development Partnership in the United States: A Case Study in Collaboration

Richard Gardner and Ron Shaffer

Introduction

Rural areas have been characterized as being cooperative and collaborative as they face the fortunes of nature and external markets. While this characterization may not be totally appropriate, the idea of friends and neighbors coming together to help raise a barn is firmly etched in the history of rural America. This history is representative of an ambitious experiment in collaborative governance called the National Rural Development Partnership (NRDP). It was conceived before the word 'collaboration'[1] became so well used. It predates the popular concepts of 'Reinventing Government' or 'New Governance', or 'social capital'. It addresses a major policy arena that is heavily fragmented – namely rural development (Radin *et al.* 1996).

The NRDP has won national recognition for its efforts, and its members have launched numerous award-winning projects. Yet its success is highly debated, even among supporters, and it has clearly not reached sufficient scale for its collaborative style to dominate the rural development arena. Funding problems have made the NRDP's future precarious, even as it is written into law with the passage of the 2002 farm bill.[2] Here, then, is the unfinished story of an attempt to do things differently.

What is Rural Development?

At the outset, it is important to clearly define what is meant by 'rural development'. Here rural development means any effort that adds to vitality or wealth of a small community. Both vitality and wealth are meant in the broadest possible sense. Vitality means not only adding jobs or income to a community, but also encompasses the adaptability or resiliency of a community, often embodied in its leaders and entrepreneurs. Wealth refers not simply to individual financial assets, but also the community's infrastructure, the natural resource endowment of the surrounding area, cultural and historical assets, and the collective skill and knowledge base of community residents. Several additional concepts can be added to this basic definition.

First, rural means more than agriculture and consumption uses of natural resources (forestry, mining, etc.). Agriculture and consumption uses continue to be the economic engine that drives many rural communities. However, it is essential that other sectors be developed to diversify the economies of rural areas and provide for economic stability. Rural development incorporates telecommunications, transportation, health care, manpower training, and poverty.

Second, 'rural development' means more than "economic development." Rural development is directly tied to education and workforce development, to accessible health care, to affordable housing, to telecommunications and technology, to transportation, and so forth.

Third, rural development is not a government program. Government must be involved, but rural development cannot be seen as solely the purview of the federal and state departments of agriculture, but rather must be the purview of all federal and state agencies which administer policies, regulations, and programs that have an impact on rural areas.

Fourth, to be truly effective and sustainable, rural development must be community driven. While the resources of government agencies, foundations, and others are desperately needed and very welcome in rural communities, too often too many decisions related to rural communities' future are made by agencies and officials who do not live in those communities.

History of rural development policy and coordination in the USA[3]

The modern history of rural development in the United States began at the turn of the century with President Theodore Roosevelt's establishment of the Country Life Commission, just as rural areas were transforming from an agricultural to a non-agricultural base. The Commission's report called for greater coordination between federal and state efforts on behalf of rural areas and recognized the interconnected nature of the elements that comprise rural communities.

More than 60 years would pass before coordinating rural development efforts would again be the focus of Presidential attention. The Cabinet level Council for Rural Affairs, established by President Nixon in November of 1969[4] was charged with coordinating federal agencies' work in rural areas, with promoting cooperation between the federal government and other levels of government, with ensuring that rural interests would be taken into account as government policies and programs were formulated, and with engaging the private and volunteer sectors as partners with government. The work of the Council for Rural Affairs influenced the work of Congress, which included a section on rural development in the Agriculture Act of 1970.

The Rural Development Act of 1972, which has served as the "charter" for all following federal rural development programs, included a first time ever provision which directed the Secretary of Agriculture to provide leadership in formulating a nationwide rural development effort in conjunction with the states. Enactment of the

Rural Development Act of 1972 led to the establishment of the interagency Assistant Secretaries Working Group for Rural Development. Although it made some progress, the Assistant Secretaries Working Group was hampered by a lack of statutory authority. The Rural Development Policy Act of 1980 was passed, in part, to address this lack of statutory authority.

Throughout this period of history the Federal government was the dominant actor in rural development. But federal dominance appears to have waned for two distinct reasons. First, the view that the federal government, with its vast financial and staff resources, should be the leader on national problems was being replaced by concerns about size of the federal bureaucracy, its competence, and its invasiveness. The challenge of improving the economic and social conditions of rural areas was compounded by the fact that the application of the resources, knowledge, and vision of other entities – the private sector, local and tribal governments, non-profit organizations, foundations, educational institutions, and others – frequently occurred in a vacuum, without cooperation with federal and state agencies and programs. Second, the pattern of standard federal (as well as state) responses being made with little flexibility for the uniqueness of rural areas of the country and nuances of the problems was being challenged. While this "one size fits all" response eased program administrators' tasks and possibly garnered the votes needed for passage, it often forced rural areas to recast their problems into some artificial form to meet eligibility.

The report of the National Commission on Agriculture and Rural Development Policy,[5] which was established by the 1985 Farm Bill, called on the federal government to: 1) take a more comprehensive approach to rural development; 2) adopt a strategic approach to rural development; 3) promote better cooperation among organizations participating in rural development activities; 4) incorporate greater flexibility in federal policies relating to rural areas; 5) promote innovation and experimentation in building more effective rural development activities; 6) incorporate greater flexibility in federal policies relating to rural areas; and 7) promote innovation and experimentation in building more effective rural development programs and institutions.

In 1988, the National Governors' Association issued its New Alliances report,[6] which called for increased collaboration between the states and the federal government in addressing rural needs. A Working Group on Rural Development was established by the President's Economic Policy Council in March 1989 and charged with exploring how the federal government might better address rural America's problems. From the beginning, the Working Group was an interagency body, including representation from the Departments of Agriculture, Commerce, Defense, Education, Health and Human Services, Housing and Urban Development, Interior, Labor, Transportation, Treasury, and Veterans Affairs, the Council of Economic Advisors, the Environmental Protection Agency, the Office of Management and Budget, and the Small Business Administration. The Working Group issued its report, 'Rural Economic Development for the 1990s: A Presidential Initiative', in January of 1990.That called for 'Improve coordination of all federal, state, and local government rural development efforts';[7] and 'Improve the support and training of local community leaders in the development process'.

Creation of NRDP and early history[8]

This brief history leading up to the Partnership points out several important aspects of the Partnership that should not be overlooked. First, the Partnership was the culmination of at least two years of internal discussion within the federal government and built on discussions started in the USDA in late 1988. This laid substantial groundwork within the federal bureaucracy to support the effort. Second, the Secretary of Agriculture, who has legal responsibility for federal rural development activities,[9] had an appreciation for the breadth of the issue, and gave the effort more than tepid support. Third, the Partnership gave recognition to the activities of the private sector, but also involved state, local, and tribal governments. The federal government indicated its willingness to join with these other partners in the formation of State Rural Development Councils (SRDCs) for the expressed purpose of building state specific rural development strategies.

The NRDP was proposed to increase efficiency of existing rural programs through greater interagency coordination, and intergovernmental collaboration. The Working Group's report was proposed by the Bush, Sr. Administration as part of an alternative six-point executive plan to a more expensive $1 billion proposal by Rep. Glenn English in the 1990 farm bill. The implementation of the Partnership became a task for the Deputy Under Secretary for Rural Development and Small Communities in the USDA and a small staff housed in the then Presidential Initiative Office (PIO), then National Initiative Office (NIO) that eventually became the National Partnership Office (NPO). Being housed at this level permitted greater leverage in spanning across various agency and departmental lines. The NRDC subordinated USDA's visibility and elevated the input of others [Hill, 1991; Lovan and Reid, 1993]. A sense of the breadth of this involvement is displayed (Table 4.1) in the number of federal departments committing staff and funds to the Partnership. The NPO and NRDC became the operational and policy groups respectively for the SRDC portion of the Partnership. The mission statements for the Partnership and NRDC listed in Table 4.2 give some sense of the operating style and desired outcomes.

Table 4.1 National Participants in National Partnership on Rural America

FEDERAL GOVERNMENT PARTICIPANTS

Appalachian Regional Commission
- Congressional Relations

Government Accounting Office

National Endowment for the Arts Outreach Office

Office of Management and Budget

Tennessee Valley Authority
- Community Resource Development

U.S. Department of Agriculture
- Agricultural Marketing Service
- Cooperative State Research Service
- Economic Research Service
- Extension Service
- Farmers Home Administration
- Forest Service
- National Agricultural Library
- Rural Development Administration
- Rural Electrification Administration
- Soil Conservation Service

U.S. Department of Commerce
- Economic Development Administration
- Minority Business Development
- Office of Legislative and Intergovernmental Affairs
- U.S. Travel and Tourism Administration

U.S. Department of Defense
- Office of Economic Adjustment
- U.S. Army Corps of Engineers

U.S. Department of Education
- Office of Vocational Education

U.S. Department of Health and Human Services
- Office of the Secretary/Intergovernmental Affairs
- Office of Rural Health Policy

U.S. Department of Housing and Urban Development
- Office of Block Grant Assistance

U.S. Department of the Interior
- Office of Program Analysis
- Bureau of Indian Affairs

U.S. Department of Labor
- Employment and Training Administration

U.S. Department of Transportation
- Office of the Secretary
- Federal Highway Administration

U.S. Department of the Treasury
- Office of Economic Policy
- Financial Management Services

U.S. Department of Veterans Affairs
- Intergovernmental Affairs

U.S. Environmental Protection Agency
- Water Policy Office
- Office of Administrator/State and Local Relations

U.S. Small Business Administration
- Office of Business Development
  - Office of Rural Affairs
- Office of Advocacy
- Development Company Branch

STATE AND LOCAL GOVERNMENT PARTICIPANTS

- National Association of Counties
- National Association of Development Organizations
- National Association of Regional Councils
- National Association of Towns and Townships
- National Governors' Association
- National League of Cities

PRIVATE SECTOR PARTICIPANTS

- American Bankers Association
- Independent Bankers Association of America
- Rural Coalition

Table 4.2 Mission Statement

- The mission of the National Rural Development Partnership is to contribute to the vitality of the Nation by strengthening the ability of all rural Americans to participate in determining their futures.

- The mission of the National Rural Development Council is to engage federal agencies and private and public interest groups to support rural communities by building collaborative partnerships and leveraging knowledge and resources in focused rural development effects through State Rural Development Councils.[10]

Source: NRDC, 1994; Sanderson, 1993.

Table 4.3 National Rural Development Partnership Guiding Principles

- That SRDCs shall include representation from all agencies, organizations, and individuals who are dedicated to enhancing economic and social opportunities for rural citizens.
- That SRDCs should have the support and participation of the governors of the states in which the SRDC operate.
- That participation in SRDC decisions and operations shall be open to all who wish to participate.
- That SRDCs shall represent the diversity of states' rural populations and geography.
- That all who participate in SRDCs shall have a voice in SRDCs decisions and that the SRDCs shall be free from control of any single individual, organization, agency, sector, political party, or interest group.
- That SRDCs shall provide open, neutral environments where all participants' voices can be heard and where major decisions are reached, whenever possible, through consensus.
- That SRDCs shall receive financial and other support from diverse sources.
- That each SRDC shall retain an executive director who will manage the operation of the SRDC; that the executive director shall be hired, supervised, and if necessary, removed through the collective participation of all of the SRDCs partners, without any one partner dictating or dominating the decision-making process; that the executive director shall be a senior-level individual with the experience, knowledge, and stature to attract all those who should be participating in the SRDC and to guide those who participate in the SRDC to effective and meaningful solutions; and the executive director shall be provided with a support staff and budget adequate to meet the expectations that have been established for the SRDC.

- That all SRDSs shall participate in the nationwide network of SRDCs; attend NRDP national conferences; participate in the governance of the NRDP; and participate in NRDP committees, task forces, and other bodies, such as Partners for Rural America.

The general operating principles of the partnership (Table 4.3) were an attempt to seek the outcomes desired by rural people. This was to be accomplished by creating forums in which the federal, state, local, and tribal governments, along with non-governmental organizations and the private sector, could come together and discuss the issues that came forth from the communities. In particular, the NRDP was to try to act in a strategic fashion and reach consensus on the issues to be addressed. Since NRDP did not have any resources itself to dispense, it required that the people sitting at the table offer to leverage the resources over which they had control. It is important to remember that in this partnership no one partner was to achieve a dominant position in decision-making. This is particularly important to remember since the various partners came from a perspective of having their view be the dominant view.

NRDP Elements[11]

The NRDP (Figure 4.1) consists of a number of components, each with a distinct purpose and function intended to further the mission. These components include the State Rural Development Councils (SRDCs), the National Partnership Office (NPO), the National Rural Development Council (NRDC), which is renamed in pending federal legislation as the National Rural Development Coordinating Committee (NRDCC), and the NRDP Executive Board (EB). As the NRDP has continued to develop over the years, each of these components has had to adjust to meet new demands and constraints.

National Partnership Office. The NPO is currently the Partnership's administrative office and is housed within USDA's Rural Development Mission Area in Washington, DC. In addition to dispensing funds to the SRDCs, the NPO reviews work plans submitted by SRDCs and convenes national conferences that provide all partners with the opportunity to further strengthen the relationships upon which the Partnership is built. An executive director, deputy director, two desk officers, a budget officer, a part-time secretary, and three Truman Fellows currently staff the NPO.

National Rural Development Council. In the early years, an executive committee of senior federal agency people essentially provided overall policy leadership to the partnership. This cluster of senior level bureaucrats and public interest groups set about to implement the SRDC portion of the Partnership. Their regular Monday meetings eventually earned the sobriquet Monday Management Group (MMG).

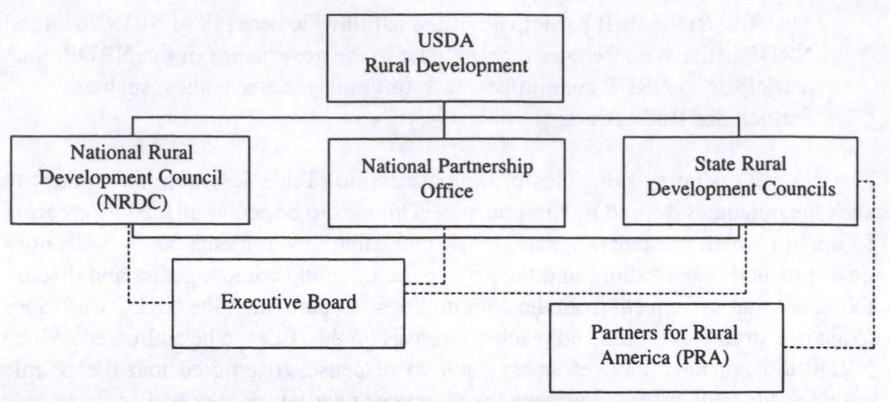

Figure 4.1 The National Rural Development Partnership

This group proved to be a key mechanism that enabled senior program managers in various federal departments and agencies to make real input into the policy management and implementation phase of the SRDC portion of the Partnership. Their networking function evolved into the National Rural Development Council, while their policy function was adopted by the Network Management Team or Executive Board.

The National Rural Development Council brings together representatives of over 40 federal agencies and national organizations to provide a uniquely rural perspective to federal policy and program development. The primary purpose of the NRDC is to serve as a conduit of information between federal agencies and the SRDCs. Specifically, the NRDC aims to identify program duplication and gaps in service to rural areas, build collaboration and coordination among federal-level rural initiatives and programs, provide input on potential consequences for rural communities during policy and regulation consultations with federal agencies and Congress, resolve unintentional federal policy or regulatory impediments to successful rural development efforts through a formal impediments process, provide a forum for the continuing dissemination of information on the status and condition of rural communities, and sustain support at the federal level for the efforts of the SRDCs. Members of the NRDC meet on a monthly basis to share policy and project developments within their agencies or organizations and to highlight significant issues facing rural America.

Executive Board. Designed to be a policy advisory body in 1995 by sharing decision-making across the partner groups, the Executive Board provides leadership to the NRDP by gathering input from members for the purpose of making recommendations, usually through consensus, to guide the Partnership. Consisting of representatives from the State Rural Development Councils, The National Rural Development Council, the National Partnership Office, and Partners for Rural America, the Executive Board often serves as the Partnership's central voice to the outside world.

State Rural Development Councils. The establishment of SRDCs was part of a comprehensive effort undertaken by the first Bush Administration to establish a more coordinated framework for developing and implementing policies and programs that have an impact on rural America. SRDCs were founded on the belief that rural citizens themselves should determine the future course of their communities; that rural community planning must be holistic in nature; that all entities with resources to invest in rural America should work more collaboratively; and that government policies should be implemented in a more flexible manner.

State Rural Development Councils exist to facilitate the establishment of collaborative partnerships among organizations and individuals who provide coordinated responses to locally identified rural needs. They provide a neutral forum at which key policy makers can address these rural needs. Frequent outcomes of this collaborative process include the leveraging of limited resources among participants, increased efficiency in the delivery of programs, mitigation of duplicative or burdensome regulatory requirements, and establishment of more flexible and responsive policies. The focus of this activity is improving the economic and social conditions of rural Americans. By the end of the 1990s, 40 SRDC had been established (Figure 4.2).

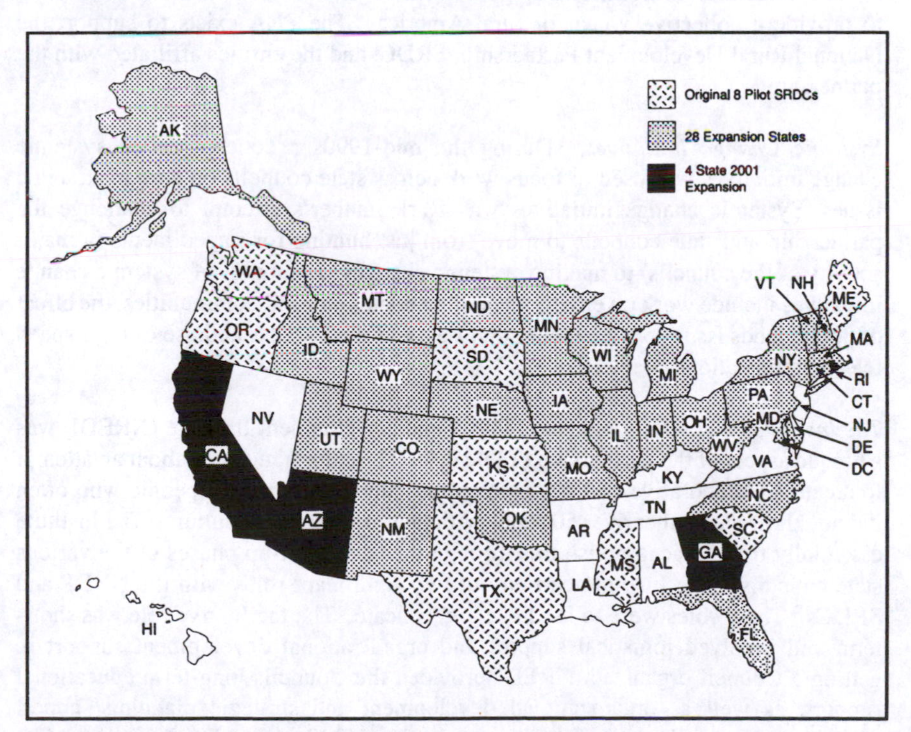

Figure 4.2 State Rural Development Council Network

Truman Scholars. One of the more interesting uses of resources was the involvement of Truman Scholars in the partnership. The NRDP created the concept of using the Truman Scholars in fellowship mode. These are recent college graduates who have chosen to work on public policy in the federal government. The partnership ended up with anywhere from three to eight Truman Scholars. Two to three of these people worked in the NPO, and the balance worked in particular federal agencies that had placed a request for them and was doing something significant for the partnership, i.e., Veterans Affairs, Health and Human Services, Transportation, etc. As NPO staff resources dwindled, these highly talented and energetic young people kept the work of the Partnership moving forward. The Truman Fellow model has since been emulated by several other Federal agencies.

Partners for Rural America. Partners for Rural America (PRA) was essentially an effort to create a unified voice for the SRDCs in approaching Congress, i.e., an attempt to overcome restrictions on lobbying. It is a 501(c) (3) and therefore is able to accept grant money from private organizations and foundations. It has only been in existence for about two years. The PRA's mission is to expand economic and social opportunities for America's rural communities and their residents; to promote equal treatment of rural America by government agencies and the private sector; and to provide a collective voice for rural America. The PRA exists to support the National Rural Development Partnership, SRDC, and the entities affiliated with the partnership.

Systemic Change Initiatives. During the mid-1990s a concept called systemic change initiatives was used to focus work across state councils on certain strategic issues. Systemic change initiatives was Karl Stauber's attempt to challenge the partnership and state councils to move from just hunting for impediments (a major activity of the councils) to making systemic changes. Examples of systemic change initiatives include work on environmental mandates on rural communities, the effect of public lands issues on small communities, and rural access and use of high-speed telecommunications.

The Institute.[12] The National Rural Economic Development Institute (NREDI) was a unique aspect of the partnership (Table 4.4). Think for a moment about an attempt to create a collaborative and strategic culture among a bunch of people who often did not know each other over 50 states with their own political culture. The Institute essentially tried to accomplish this through the early start-up phases of the various state councils. The Institute carried out two significant roles with the NRDP and SRDCs. These roles were to facilitate and educate. The facilitative role was short-term and involved logistical support and organizational development support in getting a Council organized. NREDI provided the Councils long-term educational support, as well as organizational development and strategic planning support during each Council's early formation phases (see Table 4.4). The Institute and its staff worked with the NPO staff, but did not become involved in administrative matters, e.g., evaluation of SRDC performance, budget details. The NREDI represented a subtle external intervention to help each SRDC move along on its agenda, not to determine what that agenda was nor its end point. It was important that the NREDI consultants essentially communicated the Partnership values, and tried to create a culture of collaboration and strategic action with regards to rural development.

The Institute's consultants brought a wealth of prior organizational development experience from other situations, plus other SRDCs, to aid new SRDCs through the formation stages. The task was to help each new SRDC contemplate the organizational issues it needed to address within NRDP operating principles. While no prescriptions were given, the consultants shared from their experiences and training what might work, what were some implications of choices the SRDC might make, etc. They also performed a facilitation role to help the early formation stages move more smoothly. They helped build agendas for the early meetings with the state leadership and create an organizational culture of openness and collaboration and acting strategically. As external and neutral actors, the consultants helped the SRDCs build trust and instill the collaboration mentality.

Another dimension of this early facilitative role was support for SRDC strategic decision-making. The presumption was that most members of the newly formed SRDC have had some strategic planning experiences. Rather than teach strategic planning, this support tried to build a common strategic decision-making framework among the SRDC members, to strengthen the intimate tie between their SRDC and its strategy.[13] Furthermore, it sought to create some commonality of the jargon and what the SRDCs seek to accomplish. The emphasis was to create an action/learning environment within the Council. The organizational development and strategic planning portions were viewed as seamless elements in the SRDCs' formation.

Table 4.4 National Rural Economic Development Institute

The mission of the National Rural Economic Development Institute is to assist State Rural Development Councils and their leadership to develop collaborative partnerships and adopt strategic and comprehensive approaches to rural development. To this end the Institute provides organizational development and strategic planning support to the State Rural Development Councils during their formation phases. The Institute seeks to assist State Rural Development Councils address the strategic issues facing rural people including improving the effectiveness of rural development programs.

The Institute is to provide educational and other support to assist in the formation of State Rural Development Councils within the guidelines of the National Partnership on Rural America. This support includes, but is not limited to, the following:

- Assistance in organizing State Rural Development Councils;

- Develop educational programs on comprehensive rural economic development strategies, strategic planning, new paradigms for rural development, and organizational development;
- Develop educational and technical support for executive directors and other leadership of new State Rural Development Councils; and
- Organize and conduct national programs for Council leadership and members.

Source: NREDI, 1992.

National Meetings. The second major role of the Institute was educational. This related to support of the professional development of the Executive Directors and Council leadership. This occurred via quarterly leadership meetings and two national conferences annually for council membership. The purpose of these meetings was to build the knowledge base of Council members to address their concerns in new and possibly more effective fashions. Another purpose was to enhance the sharing of SRDC experiences among the states, since this Partnership remains an unfolding experiment. These educational forums were crucial in confirming and reinforcing SRDC activities. The educational role was played out in a shared collaborative fashion rather than teacher/student. The Institute relied very heavily on people who were either in a policy making role or 'down in the trenches' to lead the learning.

It was decided that the partnership should have at least two annual meetings. In the spring, there was a meeting held in Washington, D.C. in which members of the partnership came together and essentially told their story to members of Congress. At this meeting time was also spent with one or two agencies to acquire more information about the agency and its programs and to inform it about the partnership. The second annual meeting generally occurred outside of Washington and was specifically designed to improve the learning capacity of the partnership, i.e., executive directors and executive board members. This program was built through active engagement of members of the partnership and generally had a lot of sharing by partnership members in terms of what they had learned.

With budget pressures, the NREDI was dissolved in 1998. Since then a meeting contractor has been used to help organize the logistics of national conferences. However, the loss of the technical assistance function, as well as the expertise in rural issues and institutional history, weakened the Partnership. This was especially true during the start-up of the last four SRDCs during 2000-2002.

Accomplishments

Selling the Collaborative Style. The NRDP and state rural development councils have been very successful over their history at demonstrating and helping people experience collaborative planning and implementation on a wide variety of rural issues. Twenty-five federal agencies have become involved in NRDP work, as well as dozens of state, local, tribal government agencies and private and non-profit organizations. Hundreds of individuals with rural responsibilities have been learning first hand the rewards and the pitfalls of collaborative approaches. At the risk of missing some agencies, the partnership was actively engaged with some of the following agencies in improving the delivery of their services to rural areas.[14] The agencies were: Bureau of Indian Affairs, Bureau of Land Management, Corporation for National Service, Department of Commerce, Department of Education, Department of Health and Human Services (HHS), Department of Housing and Urban Development (HUD), Department of the Interior, Department of Labor, Department of Transportation (DOT), Department of Veterans Affairs, Environmental Protection Agency (EPA), and The Small Business Administration (SBA).

Unique Funding Arrangement. A very specific example of the collaborative effort at the federal level was the unique funding arrangements of the partnership. There is currently no line item in any budget for the partnership. The partnership has survived by utilizing excess salaries and expenditure monies in the various departments. This means that most of the funding for the partnership does not flow in until late in the federal fiscal year. Over the course of its 11 years of existence, 16 departments of the federal government have funded it at times and the individual State Councils have received varying levels of support from a variety of other governmental and non-governmental organizations. Crucial initial funding at the beginning (first two years) of the Partnership was provided by: USDA/RD, DOT, HHS, Interior, EPA, HUD, SBA and Defense. One-third of an SRDC's budget must be provided from non-federal sources either in cash or in-kind contributions. Most SRDCs exceed that minimum with contributions from state, private, and foundations sources, and there have been efforts to raise the required match.

Tangible accomplishments. SRDCs have accomplished collaborative projects in every state. For instance, since the NRDP began documenting success stories on an electronic database in 1999, nearly 450 successes have been recorded. These range from small successes like workshops or community visits to major changes in program delivery like unified application forms between agencies or collaborative development of new responses to rural issues. Some state achievements have been very impressive. For example, the Idaho Rural Partnership documented 47 different collaborations in its first ten years. Illinois Rural Partners led the development of a collaborative statewide strategic plan for rural development that both state and federal agencies adopted.

The SRDCs have been able to create some very tangible outcomes. The following represent only a sampling drawn from the Partnership's 2002 website.[15] Indiana, Iowa, Maryland, Michigan, New Hampshire, Texas, Vermont and Wyoming were active in doing community assessments where teams from the council visit a community and help identify issues and resources. Then Maine, Illinois, North Carolina, Oklahoma and Pennsylvania were involved in housing programs in their states such as available resources and encouraging collaboration.

Florida and Mississippi were very involved in working with their social services agencies on such things as donating surplus items and youth-at-risk programs. Colorado, Missouri, and Washington were very instrumental in making arguments about the availability of telecommunication access in rural areas. New Mexico and Utah were working with their workforce development agencies in retraining from job loss situations. Connecticut, North Dakota, and West Virginia were very instrumental in terms of helping rural communities understand the economic development implications of health care, planning for rural health, TELEHEALTH, and access to insurance. Alaska, Minnesota, Nebraska, New York, Oregon, South Carolina, South Dakota, and Wisconsin promoted economic development through encouraging access to small business services, publicizing state initiatives, and identifying rural capital gaps. Kansas and Massachusetts were engaged in rail policy development and road guidelines. The preceding is just a partial listing, and every council was involved in multiple outcomes.

Collectively, the SRDCs and their executive directors have won numerous national awards. These include five National Performance Review Hammer Awards for reinventing government, two SBA Advocacy Awards and two HUD Best Practice Awards. In addition, the NRDP awards four Rural Impact Awards annually for path breaking work by individual SRDCs.

Learning community. An important process impact happens from both the regular meetings of SRDC members within their states and regular meetings at national levels where one conference is devoted to interactions with federal agency partners in Washington, DC, and a second conference is targeted at exchanges between states. A nested set of state and national networks has been created, with considerable development of social capital. Joint professional development training and the work on systemic change initiatives in specific issue areas between states spurs learning across agency and council boundaries. A noteworthy aspect of SRDC successes is that the best models are quickly adopted into other states through learning at the national meetings.

In addition, people tend to come to the Partnership with a subconscious perspective or paradigm of rural development. Six such perspectives are shown in Table 4.5. Both SRDC and national meetings serve to mix people of all perspectives together. As they learn and share perspectives, their view of rural development stretches and becomes more comprehensive and holistic.

Pitfalls and tensions

Even as State Rural Development Councils matured as organizations, patiently nurtured a wide variety of projects and partnerships, and developed a solid record of success, the NRDP also experienced a number of tough lessons about pitfalls to avoid and living with the tensions of conflicting interests.

Table 4.5 Perspectives on Rural Community Development

NRDP participants bring differing views of what rural community development means:

- *Economic Development* – Rural community development means creating more jobs and income. This view is held by state departments of commerce, SBA, EDA, chambers of commerce, etc. It is the perspective of economists.
- *Natural Resource Management* – Rural community development is about managing the land and natural resources surrounding the community. This view is held by federal and state natural resource agencies. It is the perspective of farmers, ranchers, foresters, miners, scientists and professional resource managers.
- *Human Service Provision* – Rural community development is about getting human services effectively provided in areas of sparse population. This view is held by those involved in workforce training, education, health care provision, and support programs. It is the perspective of educators, social workers, and health care professionals.
- *Infrastructure Financing* – Rural community development is all about making that loan or grant to a community so a tangible project can be built. This view is held by federal and state program managers in places like USDA-Rural Development, EDA, HUD's CDBG program, as well as the local grant writers and administrators who use these programs. It is the perspective of the planner and accountant.
- *Local Public Administration* – Rural community development is about balancing state and federal mandates on local government with the needs of local constituents. This view is held by many county commissioners and mayors.
- *Community Activist* – Rural community development is about empowering the disenfranchised, advocating for social change, and reducing rural poverty. This view is held by many leaders of non-profit organizations and political activists. It is the perspective of a political scientist.

Surviving the Test of Time. As with any new policy initiative or any new organization, sustaining energy and interest among the partners is a challenge. A relatively benign manifestation of the effects of time is the delegation of participation in the NRDC and SRDCs to lower levels of management. This can be good if the lower manager can give the organization more time and has more knowledge and experience in rural development, but it tends to erode political support. A larger challenge occurs when the leader who initiated the organization gives way to the next in line. The passion for the NRDP exhibited by the Bush, Sr. Administration, and Deputy Under Secretary for Rural Development Walt Hill in particular, was not present in the Clinton Administration. Because the change in administration included a party switch, there may have been increased skepticism and less priority given to initiatives from a previous administration.

Education of incoming political appointees is a related challenge. Some leaders understood the NRDP's intent, but raised legitimate questions about structure and accountability. Numerous others never appeared to truly grasp the collaborative approach, being more accustomed to the direct control afforded in line agencies. The tenure of USDA Under Secretary Jill Long Thompson was characterized by tepid public support accompanied by private expressions of doubt and a slow undermining of political and financial support.

The intergovernmental nature of the NRDP, meant that state rural development councils had to simultaneously weather transitions in administration at the state level. New governors are usually accompanied by wholesale changes in state agency directors, leading to profound change in state membership of SRDCs. Because many SRDCs were hosted by state agencies, new relationships and appreciation for the SRDC had to be quickly forged. Where party switches occurred, transitions were again more difficult. For instance, a Democratic governor housed the Idaho Rural Partnership in the Governor's budget office on its creation, where the administrator thought the interagency view would match well with IRP's partnership mission. The election of a Republican governor and two new budget office heads led to questions about the strategic importance of IRP to the budget office and an eventual move to the Idaho Department of Labor. While not antagonistic, the change in office placement caused subtle changes in how IRP was perceived. Variations of this scenario played in every state over time, with levels of success ranging from the collapse of the SRDC in Louisiana to the passage of enabling legislation in states like Washington.

Walking the Political Tightrope. State Rural Development Councils are created with the consent of the Governor, even if they are formed as a non-profit. It takes a shared act of will by an SRDC Board and the NPO to prevent it from becoming a partisan institution or from having its executive director position viewed as a patronage hire. Ideally, the tensions between political parties and between local, tribal, state, federal and private interests suspend the SRDC in a sort of neutral equilibrium. Not all states have been successful in maintaining this objective stance. A few states, such as Louisiana, Mississippi, and New Mexico, began as explicit partisan operations. Generally, they failed to become effective councils until adjustments were made to run them more neutrally.

Staying on the Policy Agenda. SRDCs were founded on the belief that rural citizens themselves should determine the future course of their communities; that rural community planning must be holistic in nature; that all entities with resources to invest in rural America should work more collaboratively; and that government policies should be implemented in a more flexible manner. Especially disappointing has been the virtual failure of efforts to better coordinate the policies and programs of federal agencies that have an impact on rural areas. Similarly, few states have made meaningful progress with efforts to formally coordinate their agencies' policies and programs that impact rural areas within their boundaries. SRDCs have been able to achieve limited coordination on a voluntary basis as partners learn the roles of others and clarify their missions. Only a few SRDCs, such as Nebraska, are endowed with the ability to set overarching state rural policy.

In addition, SRDCs face continual temptations to stray from their stated mission of fostering new collaborations around rural issues. Members may become so interested in one issue that the SRDC becomes a lead agency on that topic and takes on programmatic responsibilities. Also they may become so burdened repeating past successes that the openness to new collaborations or engaging new partners is lost.

Selling the Collaborative Style. Collaboration has many enemies, most of which are acting subconsciously. Remember that collaboration means "working with the enemy," especially to members of the World War II generation. Collaboration can be a challenge to an individual manager's or agency's sphere of control, because at collaboration's core lies a sharing of resources and decision- making. It requires people who are secure about their program's position, people who can see the benefit to their image that will come from sharing selflessly for higher purpose, and people who see a collaborative approach as a way to share the risk. For them collaboration is a way to try the idea they have long dreamed of, or the project that sits on the marginal edge of their mission area. For some, collaboration may be a way to advance change within their organization.

Avoiding Turf Battles. People representing organizations threatened by collaboration react by downplaying the potential benefits of a partnership approach and insisting their agency or program already addresses the issue sufficiently. They may withhold information or support, or even actively work against the collaboration. They prefer the status quo. In some states, SRDCs have not been formed because of such perceived threats, even though capacity for rural development may be very low. Organizations that live on grant administration seem especially prone to feeling threatened. The perception of a threat often seems to flow from an assumption that all organizations view rural development the way they do, e.g. as job creation or the construction of large infrastructure projects. (See Table 4.5.)

Accountability. SRDCs are unique organizations. They are charged with helping to improve economic and social conditions for rural citizens. However, unlike traditional government agencies or private sector entities, SRDCs are not expected to administer programs that result in tangible outputs, such as jobs, houses, bridges, immunizations and so forth. Instead, they were created to facilitate increased collaboration among those with the resources and responsibility to produce those tangible outputs. The SRDCs challenge is to not become bogged down in the administration of collaborative projects, but to remain open and enthusiastic to new collaborative proposals.

SRDCs exist in a world that is accustomed to measuring the degree to which entities are successful by counting the number of tangible outcomes they produce. To an extent, the success or failure of SRDCs can be measured by the success or failure of the entities that take advantage of the collaboration SRDCs facilitate. But this is not, by any means, a wholly satisfactory method of measuring SRDCs' success or failure. In the mid1990s the work of Cornelia Flora and other rural researchers[16] on social capital helped articulate the value of improved working relationships in community development. The creation of social capital helps explain the value of time-consuming collaborative processes. It addressed the critics who said, 'All they do is hold meetings'.

When SRDCs were established in the early 1990s, they were required to produce strategic plans that addressed rural needs found in either their own rural assessment or that of another approved entity. Quarterly reports or copies of quarterly newsletters were required, as well as either an annual report or annual newsletter. By 1996, however, both the NRDP and the political environment had changed. With most SRDCs now up and running, many within the Partnership felt that SRDCs needed a concrete and specific blueprint for achieving the goals found in their strategic plans. Meanwhile, Congress had passed the Government Performance Results Act, ushering in a new era of outcomes-oriented program assessments. In response, the NRDP[17] introduced The Outcomes Framework, and more specific work plans replaced strategic plans. Quarterly reports gave way to a mandated six success stories from each SRDC each year. Accountability and the measurement of SRDC successes became one of Under Secretary Jill Long Thompson's main criticisms in her testimony to Congress.

As this chapter is written, USDA has released a proposed SRDC Accountability System[18] that would place many of the Task Force's recommendations into regulation. These include requiring SRDCs to be structured as a non-profit or independent state agency with a diverse and neutral Governing Board. Each SRDC will be required to perform regular rural assessments, act as a rural information clearinghouse, facilitate at least four collaborations annually around both federally prescribed and locally determined issues, conduct workshops and meetings, provide a work plan, annual report, and website, conduct a self-assessment, and increase the non-federal funding share to 50 per cent. A procedure is to be established for non-performing councils to be peer reviewed. In return, the federal government may commit to increasing SRDC funding by half to $180,000 per council, expanding councils to all 50 states, creating a national web-based database to house success stories, and increasing NPO staff for more effective oversight and interaction with SRDCs.

While we clearly think SRDCs need to be accountable, we fear that much of the creativity and entrepreneurship expressed in the first ten years of the Partnership are going to be squeezed out and stifled by the bureaucratic straight jacket. Granted, most SRDCs can meet or already do meet most of these. It's just the idea that now everyone must meet all of them that represents the creeping bureaucratization.

Executive Director Dilemma. The performance of the executive director is key to the success of a state rural development council, yet the job is exceedingly difficult in all its aspects. In addition to familiarity with a wide range of rural development issues and players, the executive director must have many of the emotional intelligence competencies and shift smoothly between the different leadership styles described by Goleman, et.al (2002).[19] In addition, the executive director must be a competent program manager and grant administrator. Here are some tensions that rule the life of an executive director of an SRDC.

Sovereign tensions between levels of government: there are natural tensions between federal, state (and their county, city, and special district creations), and tribal governments over who has sovereign authority in various policy areas. The tension between state and federal government is made worse where the large tracts of federally owned land occur in the West.

Tension between NPO and SRDC Board: the National Partnership Office has interests in seeing measurable results from investment and in engaging on issues relating to the federal programs. SRDC Boards are responding to a different set of interests aimed at their perception of the rural priorities in their state and in satisfying their important rural stakeholders. Opportunities may arise suddenly from either national or state levels. A tug of war for the executive director's time and loyalty ensues.

Tension between operating for maximum effect at wholesale level of the organizations that serve rural communities versus maintaining street credibility by conducting activities at local level and being present throughout the rural parts of the state: the executive director must maintain an extensive, almost bewildering network of relationships from mayor and county commissioners to legislators, tribal leaders, nonprofit activists, community and business leaders, federal and state agency managers.

Tension between placing a priority on keeping key political leaders happy versus focusing on accomplishing projects that help rural areas: by playing to the power, one can keep the job regardless of actual accomplishment. Those who believe 'by your works, you shall be known' run the risk of political ambush by partisans or the small-minded.

The tension between accepting programmatic responsibilities in order to diversify and stabilize SRDC income and remaining true to a mission of starting collaborations: many SRDCs are offered the chance to administer programs on a fee basis. Unless this funds the hiring of additional staff, the executive director may be less able to act as a catalyst for new partnerships.

Adjusting the model

The result of the challenges and tensions listed above was a gradually more difficult process of securing funding from discretionary contributions by federal agencies. The Balanced Budget Amendment and an Executive Order mandating reductions in budget by all programs over the years of the Clinton Administration heightened these funding problems. At the beginning of the NRDP's history, the National Partnership Office had cash reserves to forward fund SRDCs a year at a time. As budgets tightened, that cash flow reserve gradually shrank until the SRDC cooperative agreements were being amended to provide only six weeks of additional funding. As this paper is written SRDCs are three months arrears in funding.

In July 1995, USDA Under Secretary Jill Long Thompson canceled the National SRDC Leadership Conference, scheduled that year for Ketchum, Idaho, at the last minute. The unilateral nature of this decision raised fresh questions about the nature of NRDP governance and its ability to live by its operating principles of collaboration. The decision also clearly signaled that the support of the Clinton Administration for this Bush initiative was lukewarm at best.

Budget pressure and the perceived lessening of political support caused the NRDP to consider options for structural change at its March 1998 meeting in Washington, DC. Five options were considered, with the Idaho delegation crafting a sixth option on the spot. The group chose by consensus the least risky option of creating a national non-profit organization that could represent the SRDCs, seek new sources of funding, and perhaps act as a fiscal pass-through agent for existing federal funds. Thus, Partners for Rural America was born.

At the same 1998 meeting the group agreed to explore the sixth option of engaging a missing partner group – Congress – with authorizing legislation carefully designed to address several administrative and funding problems. Many members of the NRDP were fearful of this approach, so it was carefully debated within the NRDP. Finally, on March 16, 2000, Senator Larry Craig, Chair of the Subcommittee on Forestry, Conservation, and Rural Development of the Senate Committee on Agriculture, conducted an oversight hearing of the NRDP, after which he agreed to sponsor enabling legislation. A consensus bill was developed in 2000 with input across the NRDP membership. This legislation was introduced at the end of the 106th Congress in October of 2000 and was reintroduced during the 107th Congress in 2001. The Congressional Rural Caucus Chairs Jo Ann Emerson and Eva Clayton developed a parallel bill in the House.

The bills were rolled into the rural title of farm bill re-authorization, and were passed into law in 2002.[20] Passage would not have occurred without the coalition built by the education efforts of Partners for Rural America (thus proving the worth of a more nimble, non-profit arm of the NRDP). The legislation authorizes $10 million in funding within USDA, encourages support from other federal agencies as well, and authorizes gifting authority. This represents a retreat from the unique collaborative funding arrangement and a recognition of the realities of traditional funding within the federal framework. It recognizes the diverse and nonpartisan membership of SRDCs, eliminates a legal impediment to full participation in SRDCs by federal employees, recognizes the NRDC as the National Rural Development Coordinating Committee and the NRDP Executive Board as a Governing Panel, gives SRDCs specific duties, and authorizes a national intermediary to handle administrative and technical assistance functions for the SRDCs.

A factor in securing passage of enabling legislation was the way the NRDP addressed criticisms over accountability. A Strategic Assessment Task Force[21] was organized in 2001 with the leadership of Partners for Rural America. It looked at issues of data, outcomes and accountability, rural policy/program coordination, governance, and finance. The recommendations contained in this honest, self-appraisal convinced many skeptics that the Partnership was willing to address weaknesses while remaining true to its operating principles.

Conclusions

Collaboration is defined by some as an unnatural act between consenting adults. An older definition from World War II is working with the enemy. These say a lot. It is something that you have to consent to, because the existing order is to parse issues into narrow fragments of authority, to act unilaterally, and to claim success as your own. It is an unnatural act for agencies and the public/private sector to create a culture of working together and asking those affected what they want to do. This requires flexibility and imagination in the response. Effective collaboration does indeed require listening to and working with those who hold different perspectives and may historically be seen as the opponents.

While the work accomplished by NRDP pales in relation to that still needed in rural America, it is worth reflecting on the progress made:

- The SRDCs are becoming forums for actively rethinking how problems are defined and solutions identified;
- The SRDCs are performing an ombudsman's role in articulating perspectives of rural people before agencies that are not fully sensitive to the effect of their programs and current rules on rural people;
- The SRDCs have made progress in breaking down geographical and jurisdictional barriers and often have created stronger working relationships between private and public sectors that transcend established lines of authority.
- The SRDCs and NRDP's bi-partisan nature appears to have survived changes in national and state administrations;
- The reduction of regulatory constrictions, while insufficient justification for the effort, has created visible and obvious early successes that have fueled enthusiasm; and
- The lack of an externally imposed agenda of tasks and schedule has increased sense of ownership and commitment to the SRDC process.

The jury is still out on the National Rural Development Partnership. Even as enabling legislation goes into effect, a financial crisis threatens the SRDCs. Success remains uncertain. Yet this case study illustrates many valuable lessons learned in the trenches of implementation including:

- Build on past experiences and staff;
- Mix in new perspectives; diversity improves the quality of responses.
- Keep the agenda open and build trust/collaboration with small successes, so new ideas get a full hearing;
- Fostering collaborations takes time. There is a valuable service performed by collaboration specialists like SRDCs.
- Collaboration creates social capital, which can be invested to tackle issues of greater strategic importance.
- The opportunities for collaboration are numerous and the variety of possible responses is limited only by the group's creativity.
- Don't overlook the need for political muscle, but don't overuse it;
- Intergovernmental, private-public partnerships have a difficult task in balancing all the political interests of various partners. Governing Boards must be proactively supportive and effectively sell the collaborative role;
- Find and nurture as many allies as possible;
- Support for the efforts of discovering new paradigms must include both intellectual exchange and pragmatic administrative responses that reinforce the new way of doing business; and
- Keep the long-term strategic issues in mind while you celebrate the short-term successes of removing impediments and program/project responses.

References

Bryson, J. (1988) *Strategic Planning for Public and Nonprofit Organizations*, San Francisco: Jossey Bass.

Flora, C. and J. Flora (1993) 'Entrepreneurial Social Infrastructure: A Necessary Ingredient', *Annals American Association of Political and Social Sciences*, 529 (September), pp. 48-58.

Freshwater, D. (1991) 'The Historical Context of Federal Rural Development Policy,' *Western Wire*, Spring, Western Rural Development Center.

Gardner, R. (1994) *Developing Collaboration in Rural Policy: Lessons from a State Rural Development Council*, National Public Policy Education Conference, Boise, Idaho, September 21, 1994.

General Accounting Office (1994) *Rural Development: Patchwork of Federal Programs Needs to be Reappraised*,Washington, D.C.

Goleman, D., Boyatzis, R. and A. McKee (2002) *Primal Leadership: Realizing the Power of Emotional Intelligence*, Boston, MA: Harvard Business School Press.

Lovan, W. and J. Reid (1993) *Reinventing Government in Rural America*, Discussion paper, Washington DC: Rural Development Administration, USDA, April.

National Commission on Agriculture and Rural Development Policy (1990) *Future Directions in Rural Development Policy: Findings and Recommendations*, Washington DC: USDA, December.

National Governors' Association (1988) *New Alliances for Rural America*, Report of the Task Force on Rural Development, Washington D.C.

National Initiative Office (1991) *Expected Outputs from State Rural Development Councils*, Washington D.C.: Rural Development Administration, USDA, October.

National Partnership Office (2002) *Draft Proposed SRDC Accountability System*,. Washington, DC.: USDA.

National Partnership Office and Partners for Rural America (2001) *National Rural Development Partnership Strategic Assessment Task Force Final Report*, Washington, D.C., USDA.

Osbourn, S. (1988) *Rural Policy in the United States: A History*, Washington DC: Congressional Research Service, Library of Congress.

Radin, B., Agranoff, R., Bowman, A., Buntz, C. Ott, S., Romzek, B. and R. Wilson (1996) *New Governance for Rural America: Creating Intergovernmental Partnership*, Lawrence: University of Kansas Press.

Rasmussen, W. (1985) '90 Years of Rural Development Programs', *Rural Development Perspectives*, Vol. 1, No. 3.

Sanderson, D. (1993) *Adjusting the Course: Re-defining the National Council's Mission and Roles: A Report of the October 1993 Retreats of the National Rural Development Council*, Washington, D.C.: NPO-USDA, December.

Shaffer, R. (1994) 'State Rural Development Councils and the National Partnership for Rural Development', *New Governance Discussion Paper Series*, No. 7, Madison, WI: National Rural Development Inst (revised Feb 1996).

Walter, E. (1991) *Making Rural Policy for the 1990s and Beyond: A Federal Government View*, Paper at 1992 Agricultural Outlook Conference, Washington D.C.: US Dept of Agriculture, December.

Notes

1 What is meant by collaboration? It is more than co-operation, where authorities inform others of what they plan to do anyway. It is more than co-ordination, where authorities share information and alter what they plan to do anyway to mesh with what others intend to do anyway. Collaboration involves committing decision-making authority and resources to a group of stakeholders with a shared interest in taking action on an issue. It is the process of building trust and respect in order to achieve a common objective from uncommon perspectives [Gardner, 1994].

2 SEC. 6021. NATIONAL RURAL DEVELOPMENT PARTNERSHIP. Subtitle D of the Consolidated Farm and Rural Development Act (7 U.S.C. 1981 et seq.) (as amended by section 5321) is amended by adding at the end the following: SEC. 378. NATIONAL RURAL DEVELOPMENT PARTNERSHIP.

3 This section drawn from Freshwater [1991], Osbourn [1988], National Partnership Office and Partners for Rural America [2001], and Rasmussen [1985].

4 Executive Order 11493, Council for Rural Affairs, November 13, 1969.

5 National Commission on Agriculture and Rural Development Policy, 1990.

6 National Governors' Association, New Alliances for Rural America: Report of the Task Force on Rural Development, (Washington, DC, 1988).
7 If there was ever a question about the need for better coordination of federal programs that have an impact on rural areas, that question was put to rest by a 1994 study of the General Accounting Office (GAO) that identified over 800 such programs. In just the area of infrastructure development, the report identified 84 programs administered by 13 different agencies.
8 Drawn from Shaffer [1994].
9 1972 Rural Development Act.
10 This national mission and goal statement applies to the NRDC only. Individual state rural development councils adopt their own state-level mission statements.
11 Drawn from National Partnership Office and Partners for Rural America, December 2001 and Shaffer [1994].
12 Drawn from Shaffer [1994].
13 Bryson [1988, p. 59] notes strategy is a pattern of purposes, policies, programs, actions, decisions or resource allocations that define what an organization is, what it does, and why it does it. Effective strategy formulation and implementation processes link rhetoric, choices, and actions into a coherent and consistent pattern across levels, functions, and time. Some key words are – rhetoric – what people say; choices – what people decide and are willing to pay for; actions – what people do.
14 http://www.rurdev.usda.gov/nrdp/agency.html
15 http://www.scicomminc.com/nrdpsuccess/new_listing_priv_n.asp
16 Flora, C & J Flora (1993).
17 See National Partnership Office, 1996.
18 See National Partnership Office, May 30, 2002 for more detail.
19 Goleman, Daniel, Richard Boyatzis, and Annie McKee, 2002.
20 SEC. 6021. NATIONAL RURAL DEVELOPMENT PARTNERSHIP. Subtitle D of the Consolidated Farm and Rural Development Act (7 U.S.C. 1981 et seq.).
21 See National Partnership Office, May 30, 2002 for more details.

Chapter 5

Governance and Community Engagement: The Australian Experience

Jim Cavaye

Introduction

The role of government in Australia is changing in order to stay the same. Government agencies are adopting new processes of community engagement, thus reconnecting government with an unchanging purpose of democracy relating to the fostering of citizenship. This chapter outlines the insights from community engagement experiences, suggests new approaches, and describes the actions that would better support citizen participation.

The notion of community engagement and whole of government coordination has broad support. The difficulty has been in how to achieve it. Rather than seeking the best model of engagement, I contend that a more flexible system of governance is needed to allow government agencies to better manage the dilemmas and trade offs inherent in true community engagement. This calls for changes in policy, structures and practice. Most importantly it demands changes in the assumptions, values and culture of public administration.

Before proceeding I must place government interaction with communities in context. First, government agencies and individual communities do not relate to each other exclusively. Communities engage in a complex network of interaction, including private enterprise, community groups, and individuals – as well as public agencies. Second, there is also no one government or community. I refer to government as mainly state and federal government departments that interact with communities. As such, government consists of a range of public agencies with different roles and cultures. Communities are also diverse, consisting of a wide range of sectors, groups and individuals with differing perceptions, interests and interactions with government. A third context lies in the potential influence of government. While government certainly affects communities and vice versa, there is a limit to the engagement and influence government can have with communities. What then do we mean by community engagement? What has led to such a focus on citizen participation in Australia?

What is community engagement?

Community engagement is mutual communication and deliberation that occurs between government and citizens. It allows citizens and government to participate mutually in the formulation of policy and the provision of government services, but often with the final responsibility lying with the elected government (OECD, 2001). Ideally it links government action with community action to progress community and government goals. Various arrangements, structures and processes can mediate this interaction.

Community engagement necessarily means participation with a community of people, rather than an individual citizen. This means that engagement arrangements need to incorporate the diversity and dynamics of communities, issues of community representation and power, and the potentially conflicting goals of sub-communities. Government and communities can engage each other at several different levels.

Levels of engagement

Figure 5.1 shows the many degrees to which government and communities can engage with each other. The most basic form of citizen participation is simply casting a vote – an act of participation in government that is compulsory in Australia. Community members become slightly more involved when government informs the community of decisions. Citizens participate passively in a one way relationship with government which informs citizens of decisions and disseminates information. Consultation involves two way communication between government and communities but only in providing feedback to government. Often focused on a specific issue or proposal, government defines issues and controls decisions. Citizens comment on the proposal or issue, or provide information and opinion.

Structured community involvement entails advisory committees or representative panels that mediate community input. People may join with government on specific projects or other forms of formal involvement. Material incentives (such as a funded project) encourage citizen participation or people may contribute their time and resources. Government and communities make some decisions jointly, but often project goals are pre-determined. Community partnership involves government and communities in joint decision-making, shared leadership and common goals. People participate equally and fully in a joint learning process. Involvement is natural and self-sustaining. Finally, government agencies can act to facilitate community-led action. Communities self mobilize, controlling decisions, resources and actions. Government can help facilitate the community process at the invitation of the community.

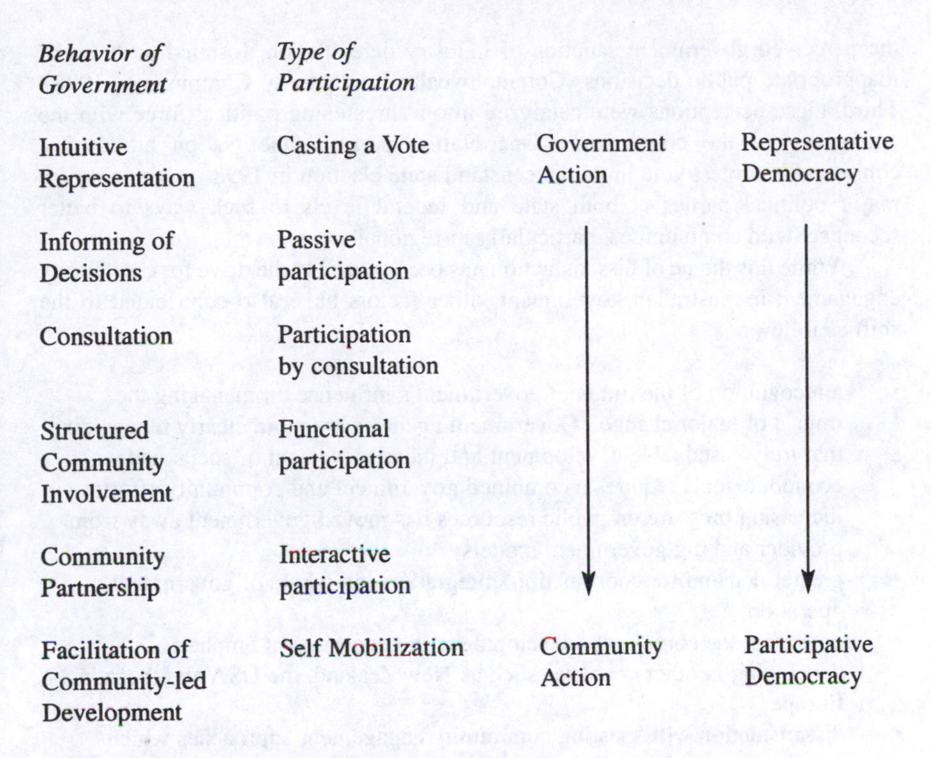

Behavior of Government
Type of Participation
Intuitive Representation
Casting a Vote
Government Action
Representative Democracy
Informing of Decisions
Passive participation
Consultation
Participation by consultation
Structured Community Involvement
Functional participation
Community Partnership
Interactive participation
Facilitation of Community-led Development
Self Mobilization
Community Action
Participative Democracy

Source: adapted from Cavaye (1999), Pretty (1995), Arnstein (1969).

Figure 5.1 A Spectrum of Government Interaction with Communities and Forms of Participation

A shift to engage communities

Progressing engagement beyond consultation, and developing more effective engagement processes has been a key issue for government in Australia since the mid 1990s. Several factors have combined to create this emphasis. First, accumulating economic and social changes continue to have enormous impact, particularly in rural and regional areas. People's expectations of an appropriate government response have generally not been met, creating popular disillusionment with government generally. Federal and state governments, in particular, were seen as not listening and out of touch with citizens. Second, government itself was seen to have contributed to regional decline. The widespread withdrawal of government services and the rationalization of facilities during the 1990s were a major source of dissatisfaction. Policies such as National Competition Policy, and in some states, the perceived government sanction of industry deregulation, fostered attitudes of inappropriate public decisions (Commonwealth Productivity Commission, 1998). Third, these perceptions were catalyzed into a threatening political force with the emergence of the conservative One Nation party. One Nation attracted a considerable protest vote in the Queensland state election in 1998, galvanizing the major political parties at both state and federal levels to seek ways to better reconnect with communities, particularly in regional areas.

While this theme of dissatisfaction has been central to the drive for community engagement in Australian government, other factors have also contributed to the shift as follows:

- a recognition of the limits of government's influence on managing the impact of major change. Government agencies are more clearly recognizing that truly sustainable development and the achievement of social and economic goals requires a combined government and community effort;
- increasing pressure on public resources has moved government away from provider and big government models;
- greater demand for coordination, integration and whole of government operation;
- observing the community development and engagement emphasis developing in other countries such as New Zealand, the USA and those in Europe;
- dissatisfaction with existing community engagement approaches which largely amounted to uncoordinated traditional consultation activities.

Finally, one of the most crucial drivers has been greater community expectations of input to government decisions and processes. This is partly borne out of dissatisfaction with existing approaches. However, I contend that it also derives from a general maturity of democracy in Australia where citizens are expecting more participation with government, and government culture appears generally less accepting of unilateral decision-making. The willingness of citizens to trust government has been declining for decades (Bush, 2001), and anecdotally there is much evidence of greater citizen expectations of input to government decision-making.

The recent focus on community engagement represents part of an ongoing evolution in the role of the state, and the dominant approach taken to achieve the role of government in Australia and in western democracies generally (Figure 5.2). From the basic Weberian views of democracy, several models of government provide a background to current approaches to community engagement. Kenny (1999) describes several models of government such as the Independent State, which relies on representative democracy and elected officials making good public decisions on behalf of citizens. The Instrumental State aims to produce policy to influence community outcomes. The Interlocking State depends on the integration of policy, service delivery and disciplines to deliver coherent public benefits.

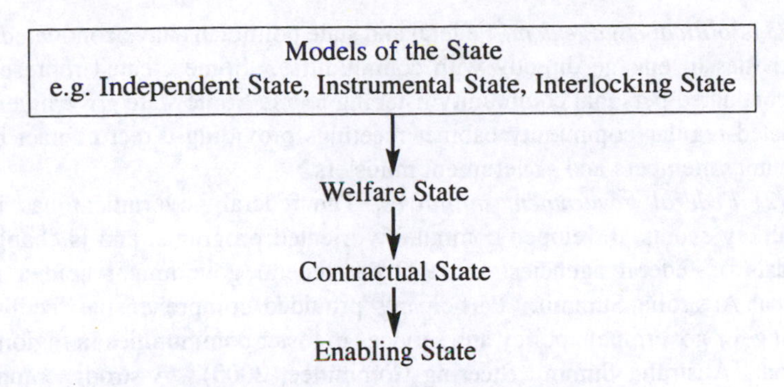

Figure 5.2 A General Progression in Government Forms

These basic conceptions of government provide a context for the emergence of the Welfare State of the 1960s and 1970s with the development of major social policy and infrastructure. The Contractual State emerged in the 1980s with a particular emphasis on the adoption of business principles and practice by government. Customer service, purchaser/provider models and user pays characterized this approach (Barzelay, 1992; NPR, 1993). The current focus on community engagement and citizen participation is part of the emergence of an Enabling State (Botsman and Latham, 2001). Government is attempting to build a facilitation and partnership role with communities that better enables community capacity and adds value to community outcomes. How then can government engage with communities and develop an enabling role? What has been the experience in Australia?

Community engagement in Australia

Community engagement in Australia is not new. Citizens and government already work together on a wide range of existing activities. For example, land-care and catchment management have involved extensive government and community cooperation in managing natural resources. Many volunteer and community groups work cooperatively with government departments, for example in emergency services, community policing, parents and citizens associations, and in health care. It is important to recognize that increased community engagement builds on this extensive base.

Increased emphasis on community engagement has been expressed in four main forms – politically, by federal government, through state structures and programs and through changes in delivery-level practice.

(1) Political engagement: Federal and state politicians have conducted a range of activities to engage directly with communities. Some elected representatives have conducted personal community listening tours. Some state governments have conducted regular community cabinet meetings providing direct contact between community members and government ministers.

(2) Federal government initiatives: The federal government has initiated several key events, developed community-oriented programs, and is changing the emphasis of federal agencies. In 1999 the federal government held a national Regional Australia Summit. Participants provided comprehensive feedback and planning for government policy and services to foster communities in regional areas (Regional Australia Summit Steering Committee, 2000). A strong emphasis on community empowerment identified improved delivery of government services as a key theme for change. A North Australia Summit followed in 2001.

New federal programs have focused on providing funding and support for community-based activities and fostering community partnership in service delivery. Some key initiatives have been Rural Plan within the program Advancing Australia Agriculture; the Stronger Families and Communities Strategy, the Regional Health Strategy, the Rural Communities Program, the establishment of rural transaction centres, the Regional Australia Strategy, the Regional Solutions Program, the Stronger Regions - A Stronger Australia Statement and the establishment of a Foundation for Rural and Regional Renewal. These programs are aimed at improving the delivery and impact of federal government services largely in regional areas. They consistently feature government partnership with communities, coordination and bottom-up citizen involvement.

(3) State government structures and programs: State governments have attempted to move strongly towards community engagement and fostering community-based solutions. For example, Western Australia developed a community development focus through initiatives such as Community Builders and the WA Rural Leadership Program. New South Wales focused on place management, community relationships and local integration of services through its Strengthening Local Communities Strategy. Similar initiatives have developed in other states such as the Community Capacity Building Initiative in Victoria and Community Renewal projects in Queensland.

States have also developed specific structures for community engagement. For example, The Regional Communities Program in Queensland is based on direct ministerial participation with about 20 community representatives in each of eight regional forums. A Community Engagement Division within the Department of Premier and Cabinet has a specific role in leading and implementing the government's engagement with communities.

(4) Service delivery and practice: Government is also attempting to engage with communities at the service delivery level in two main ways. First, a range of

specific service delivery projects have been based explicitly on community engagement. The Central Highlands Regional Resource Use Planning Project, Community Public Health Planning Projects, Community Renewal Projects and the Rural Partnership Program are examples of specific attempts to make community participation an integral part of service delivery. These projects represent a distinct move down the level of the engagement spectrum shown in figure 5.1. They aim to foster partnership and a greater government contribution to helping communities build their own capacity. Second, many agencies are attempting changes to existing practice. They are developing community engagement strategies and building community participation into some current projects.

Despite these changes to practice, community consultation remains the most common form of government engagement and many communities are feeling consultation fatigue. Consultation on contentious issues such as irrigation water allocation or vegetation management has highlighted conflict and impasse between government and major community stakeholders. What are the issues raised by these approaches to community engagement? What does government need to pay more attention to, and what are potential new approaches?

Clues to good community engagement

Attempts to improve community engagement have been a struggle of ideas, expectations and assumptions. Within this fluid situation, many of the experiences above have illustrated some key elements of successful community engagement. Based on largely anecdotal evidence some consistent themes are:

- local relationships and continuity of contact with individuals in government is valued by communities. Particular individuals both in government and communities play an important part in engagement;
- the extent and sophistication of leadership in both communities and government is important to the quality of engagement;
- people's motivation to participate relies on their perceived influence with government, and their passion and commitment for their community or for particular issues. People need to be confident that their view will be heard and may contribute to real change;
- engagement processes must have authority and legitimacy to influence issues and on-ground practice;
- some structures, procedures or relationships need to mediate community engagement;
- follow up to engagement events or in progressing issues is crucial to community engagement and to local confidence with government;
- community engagement needs to be closely coordinated across agencies and agencies need to avoid duplication or over-engaging;

- clear practical demonstrations of engagement help to foster further engagement – a show me, don't just tell me approach;
- power relationships and self interest in communities need to be managed. Political motives need to be managed within government;
- engagement processes must be compatible with government accountability, political expectations and performance goals.

These elements of engagement from general experience add to the elements of sound government-community engagement outlined in a publication by the Queensland Government (2001b):

- Consistent protocols across government
- Respect and open mindedness
- Timing (e.g., adequate notice and realistic timeframes)
- Adequate resources
- Accessible information
- Cultural appropriateness
- Accountability
- Partnership
- Decentralization
- Maintaining reasonable expectations

Indeed, a general framework of the critical success factors in good government-community engagement based on Australian and some international experience is summarized in Table 5.1.

Table 5.1 The Components of Government-Community Engagement

Elements of good Government-Community engagement	*Factors that support elements of community Engagement*
"Will", genuine motivation	Motivation to engage and achieve an outcome Negotiated expectations and limits
Relationships and trust	Accessibility Reciprocity Communication Consistency Continuity of contact
Leadership	Shared leadership Collaborative focus for leaders Attitudes and skills of leaders
Decision-making	Legitimacy to influence decisions A decision-making purpose for engagement
Inclusiveness	Diversity of community included Equity of opportunity to participate Processes that allow broad participation Information and awareness
Structures, procedures	Organizational arrangements Protocols Techniques and methods
Accountability	Engagement processes accountable as good practice Government accountability for outcomes from engagement Government and community with mutual obligations
Skills	Ability to manage conflict, include diversity, maintain quality communication
Satisfaction	Gauging the extent of satisfaction with engagement Managing expectations and distinguishing the process from the outcome
Follow-up, sustainability	Appropriate ongoing engagement Feedback

Source: based on Queensland Government (2001b); Bush (2001); Putnam, 1993; National Economics (1999).

Key risks

Experiences in Australia arguably reflect many of the elements of good community engagement. However, they also raise many issues and challenges. Some key risks are emerging.

(1) Changing practice but not changing assumptions: there is a risk of government embracing community engagement and a capacity extension role but attempting it with the assumptions and principles of service delivery and technical assistance. Genuine partnership requires different assumptions, values and principles to a traditional delivery approach. There are examples of traditional thinking in community engagement approaches that amount to "we are from the government and we are going to engage you", rather than valuing and investing in relationships and building true partnerships. Without new thinking, government agencies can subtly develop a mindset that supports the delivery of community engagement.

(2) Over-emphasis on structures: structures and processes that mediate communication between government and communities are crucial to engagement. However, I argue that a strong focus in Australia has been on developing structures, programs and organizational entities, with insufficient emphasis on cultural change processes within agencies to develop the principles and norms that provide a context for structures.

(3) Listening better: there is a tendency for government to limit the objective of community engagement to listening better. Clearly providing feedback to government, which has the elected legitimacy and responsibility to implement public decisions, is a key role of engagement. However, it is easy for engagement to reinforce the perception that government has responsibility for community improvement. Engagement can be seen simply as a better means for communities to tell government what their needs are and what government should do. To the contrary, true engagement is a dialogue incorporating not only what government and communities can do alone, but also how they can add value to each other. While government has ultimate responsibility for public outcomes, a truly enabling role involves government and communities in a relationship with mutual input to community outcomes.

Other risks

There are a number of other potential risks. First, the idea of building community capacity undervalues the existing often informal capacity of communities and reinforces paternal approaches to communities. Community capacity building needs to be reframed into capacity appreciation or extension, or helping local people build their community's capacity. Second, while community of place is a very important aspect of government's relationship with communities, it may create tension in engagement (Stewart-Weeks, 2000). Greater recognition of communities, sub-communities of interest and other ties need to be included in engagement strategies. Third, agencies have a clear role to provide infrastructure and services. However, this needs to be balanced with investment in relationships and leadership for effective community engagement. And fourth, it is arguable that more attention needs to be paid to the quality of engagement. Much of current engagement is based on intuitive involvement – beliefs that the more people that are involved the better, or an open meeting draws a representative sample of people. A more sophisticated approach to engagement is needed which better accounts for community dynamics and the quality of engagement.

Issues

Community engagement also raises many issues and questions:

- what are the necessary precursors for effective government – community engagement and partnership? What attitudes, organization, and relationships make communities and government agencies prepared for engagement?
- how does engagement balance debate on the major issues and policy, and the need to deal with micro community issues?
- processes need to incorporate multiple motives for engagement, including, perhaps community motives to lobby or criticize government and government's motive to showcase community engagement;
- how can communities and government enhance their own capacity to engage with each other? How can community members manage emotion, anger, power and self interest? How can government agencies develop the skills, attitudes and resources to coordinate and manage a genuine relationship with communities?
- who represents the community in engagement processes? How does government balance the legitimacy of formal leaders and community organizations with the popular legitimacy than comes from broad community participation and informal leadership?
- to what extent does effective engagement consist of participation events or ongoing involvement, or both?
- when is engagement necessary and when is it not? A strong focus on engagement may create its own culture of "when in doubt, engage". This may push engagement into situations where it is inappropriate or maintain levels of over-consultation;
- when is there enough community engagement? At what point does government consider that an appropriate level of engagement has been reached or attempted? On what issues or in what circumstances do community members want to be engaged, and when should government act without community contact;
- what practical strategies help communities and governments manage competing interests and conflict?

- in expanding community engagement, government faces a strong history of dissatisfaction in many communities. How can this be managed and how can expectations account for the fact the government is "starting behind scratch";
- how does government manage the risk involved in community engagement? Potential risks are committing government to actions it cannot fulfill, over-committing meager resources, exposing agency business to political leverage through community lobbying and possible political embarrassment;
- to what extent is engagement framed in terms of government providing better services and policy as opposed to enhancing the broader role of fostering vibrant sustainable communities;
- who has control? Can government delegate power and retain full control?
- to what extent is a comprehensive engagement policy needed, as opposed to the flexibility of more specific approaches?
- how does government ask communities how they want to be engaged?
- when should government not engage communities? In some cases, government agency contact may hinder community empowerment and development. There is substantial evidence of community groups being co-opted or community capability being suppressed by government (Piven and Cloward, 1979; Moynihan, 1969; McCloskey, 1996).

Dilemmas

Community engagement also highlights fundamental and operational dilemmas for government:

- in working with communities government agencies must embrace pluralism and diversity, often mediated by community power relationships. Yet agencies must also retain the basic democratic principle of equity – making decisions for the overall good of all and not be seen to favor one group over another. The overall public good can supersede what specific communities value. Water allocation and vegetation management are examples of trade-offs where government is struggling to balance general societal outcomes with the goals of interest groups;
- agencies are under strong pressure to be efficient with time and resources and deliver clear measurable outcomes. Yet the nature of engagement is often inefficient and the effort and investment in engagement processes is often considerable. Community Renewal projects, for example, have highlighted the time and effort involved in developing community relationships and negotiating community issues. At what point do effort and resources in community engagement substitute for the effort and resources in service delivery? To what extent are they synergistic?
- the traditional role of service delivery and direct assistance needs to coexist with an enabler role. The relatively one way relationship involved in service delivery must combine with the two way link involved in engagement;

- agencies need to balance the development of relationships and partnership with communities with the reality or perception of agency "capture";
- government agencies must retain political and public accountability for their performance. This may offer little freedom for agencies to engage communities which often involves experimentation, failure and flexibility;
- community engagement may often involve tailor-made solutions in different communities. Yet government must maintain the principle of equity and minimize precedent;
- a cornerstone of the efficiency of bureaucracy is the delineation of responsibility. Bureaucratic structure and defined responsibilities are aimed at preventing duplication and ensuring efficient delivery. Services are also necessarily differentiated to provide efficient access for citizens. Yet community engagement rarely is accepted as core business and the delineated silos of government often work against a seamless whole-of-government relationship with communities;
- community relationships and partnership involve long term commitment and continuity of contact. Yet the realities of political terms and budgets necessarily have a shorter timeframe;
- on the one hand, community engagement implies the involvement of as many citizens as possible. On the other hand, not all citizens can, or may want to be involved. To what extent then do government agencies act on the issues as expressed by the motivated few or spend effort engaging the broader community?
- agencies must balance the provision of resources and expertise from outside with the importance of maintaining genuine community ownership and self reliance;
- the attitudes and culture of communities often differ from the assumptions and norms of the government institutions. Community engagement processes need to respect and accommodate the cultural rules and expectations of each;
- engagement processes must deal with both the tangible and intangible aspects of community development. The hard issues of jobs, infrastructure and income need to be simultaneously addressed with softer community motivation, perceptions and values.

Many of these issues and dilemmas spring from the basic tension between representative and participative democracy. How then can government manage these issues and dilemmas?

Engagement governance

The Australian experience shows that there is no best way to engage government with communities. Rather than a specific model, engaged government calls for a new governance – a diverse flexible set of principles, structures and methods that can help government and community members manage dilemmas, cope with risks, experiment, and implement tailor-made approaches. The attitudes and processes in engaged governance fundamentally allow government to live with duality. It provides a way for government to manage representative ***and*** participative democracy – to not just be an enabler, but ***both*** a deliverer and enabler. It is a cultural framework that guides everyday decisions that balance basic dilemmas such as community partnership vs bureaucratic neutrality, community accountability *vs* political responsiveness, and service delivery vs empowerment.

Indeed, Gleeson and Barlow (2000) and Putnam (1993) call for new values, relationships, and assumptions in order for government to build a more engaged relationship with communities. Kingma and Falk (2000), and Stewart-Weeks (2000) argue that community engagement relies on government acting with new values and structures to help foster social capital. They argue that enhanced social capital is a key to reconnecting government with communities. OECD (1996) also emphasizes social capital, relationships and trust. Hence, engagement is not just about increasing community participation in what government does. It also fundamentally involves a shift in how government works. But this is not a revolution or another "reinvention" of government. It is an incremental change process that incorporates old and new roles, builds on existing circumstances, and allows community linkages to mature and develop rather than simply proliferate.

How might engagement governance develop?

Developing governance for community engagement involves four key aspects:

(1) Principles, values and assumptions. A clear set of values across government needs to underpin community engagement and agencies need a consistent set of operating principles.

(2) Structures, services and mechanisms that mediate two-way community engagement. Organizational structures, arrangements and processes need to support a set of relationships and collaborative networks.

(3) Practice. Grassroots-level attitudes, relationships, individual behavior, techniques and methods form an everyday *modus operandi* for community engagement.

(4) Culture. A culture of engagement needs to go hand in hand with structures and arrangements. Language, practice, relations and behavior need to culturally demonstrate and sustain good engagement practice (Kenny, 1999).

Developing these principles, structures, practice and culture involves incremental cultural change. Just as there is a risk of agencies engaging communities with traditional approaches, fostering this cultural change requires new assumptions and approaches to traditional public agency change processes. In many ways agencies are communities, with the same diversity, sub-communities and power structures as the communities they serve. Engagement challenges agencies to build their capacity as a community, and to apply the principles of community development to their own organizational change. As with communities building their own capacity, this process involves agencies in a relatively long term process that relies on internal champions and enthusiasm, fostering agency social capital, starting small, addressing the existing concerns of staff, participation, demonstration of engagement, ways in which agency staff can be involved, consistent leadership, conflict management, realistic goals, and recognition of success.

The development of engagement governance will be individual for each agency and sustained by internal drivers. A certain chemistry is required. Kingdon (1995) argues that opportunities for change in government approaches occur when problem, policy and political streams align. That is, when practical challenges coincide with political motives and policy needs. Community engagement seems to be a timely alignment of the three. Yet there is substantial inertia within agencies to community engagement. Many of the issues and dilemmas above make the cultural change to engagement governance difficult. One key issue is the change fatigue of agency staff with perhaps many staff deciding to ride out a push to engagement as yet another temporary focus. The traditional expectations of communities may also be a key source of inertia

Action

Despite the complexity and uncertainty of an incremental change process, there are some clear actions that are most likely to support community engagement in Australia.

(1) Establishing principles. High-level leadership is required to establish and sustain principles of community engagement. This would not only include the principles of community engagement but also the values that would guide an action learning process within agencies.

(2) Structures and arrangements. New forms of accountability are required to build community engagement into the performance agenda of agencies. Cavaye (1999) has proposed a tiered system of agency accountability for service delivery, the quality of engagement, and community capacity outcomes. More refinement and testing of performance indicators for community/government interaction are needed. Agencies need to be more familiar with innovative methods and techniques for community engagement. They need to be more inventive in arrangements that mediate engagement. For example, the Queensland Government (2001a, 2001b) outlines ways of creating spaces for community contact and issues of e-democracy. Other priorities for new arrangements are practical strategies to deal with competing needs, diversity and power relationships; negotiating community and government values; and whole-of-government coordination.

(3) Practice. At least four actions will enhance community engagement practice: greater training and skilling of public sector employees in community

engagement; use of new forms of participation such as coalitions or appreciative enquiry; investment in relationships; greater development and exposure of beacon projects.

Conclusion

Engaged governance is not a new policy, but rather a new politics – a new set of relationships and interactions between agencies and communities. This new politics, together with the capacity building process within agencies is dynamic and can be self-reinforcing. New values can create new policies and structures. In turn, new approaches and structures can create new relationships and trust. Agencies need the space to start working more closely with communities and by doing so they will potentially create the relationships that will help them work more closely together. At the same time, agencies may develop the expertise and confidence to manage the multiple issues and dilemmas involved. In many ways the role of government is a construct. Politicians, agency staff and citizens cast government in particular roles in communities. Engagement governance requires rethinking assumptions, structures and culture about how the work of government is constructed. The central focus is the view of government not as a provider, but as an enabler of vibrant communities. In that regard, community engagement has the potential not to challenge government, but to enhance it.

References

Arnstein, S.R. (1969) 'A ladder of citizen participation', *Journal of the American Institute of Planners*, Vol. 30, No. 4, pp. 216-24.

Barzelay, M. (1992) *Breaking Through Bureaucracy: A New Vision for Managing in Government,* Berkeley: University of California Press.

Botsman, P. and Latham, M. (2001) *The Enabling State*, Annandale: Pluto Press Australia.

Bush, R. (2001) *Theories and Models of Public Participation and Governance: Public Participation Policy and Practice in State and Local Government*, Course in Public Participatio, Brisbane: Queensland University of Technology.

Cavaye, J.M. (1999) *The Role of Government in Community Capacity Building*, Brisbane: Queensland Government, Department of Primary Industries.

Commonwealth Productivity Commission (1998) *Issues Paper – Inquiry into National Competition Policy* Commonwealth Government Canberra (obtained from www.pc.gov.au)

Gleeson, T. and Barlow, S. (2000) *Australian Values – Rural Policies.* Paper at the Australian Values, Rural Policies Symposium, Canberra, 2000, Synapse Consulting.

Kenny (1999) *Developing Communities for the Future: Community Development in Australia,* Melbourne: Nelson Publishing.

Kingdon, J.W. (1995) *Agendas, Alternatives and Public Policies*. New York: Harper Collins College Publishers.

Kingma, O. and Falk, I. (2000) *Cooperation and Tolerance: Restoring Our Economic System,* Paper at the Australian Values, Rural Policies Symposium, Canberra, 2000. Synapse Consulting.

McCloskey, M. (1996) 'The Limits of Collaboration', *Harper's Magazine*, November 1996.

Moynihan, D.P. (1969) *Maximum Feasible Misunderstanding: Community Action and the War on Poverty*, New York: The Free Press.

National Economics (1999) *State of the Regions Report*, Canberra: National Economics.

NPR (National Performance Review) (1994) *From Red Tape to Results: Creating a Government that Works Better and Costs Less*, Washington D.C.: Report of the National Performance Review, Vice President Al Gore.

OECD (1996) *Better Policies for Rural Development*, Paris: OECD.

OECD (2001) *Engaging Citizens in Policy-Making: Information, Consultation and Public Participation*, PUMA Policy Brief No 10. Paris: OECD.

Piven, F.F. and Cloward, R.A. (1979) *Poor People's Movements: Why They Succeed, How They Fail*, New York: Vintage Books.

Pretty, J.N. (1995) *Regenerating Agriculture. Policies and Practice for Sustainability and Self-Reliance*, London: Earthscan Publications Ltd.

Putnam, R.D. (1993) *Making Democracy Work: Civic Traditions in Modern Italy*, Princeton: Princeton University Press.

Queensland Government, Department of the Premier and Cabinet (2001a) *Community Engagement Division Directions Statement*, Brisbane: Queensland Government.

Queensland Government, Department of the Premier and Cabinet (2001b) *Reading Materials for "Seeking Good Practice in Community Engagement" Workshop*, Brisbane: Queensland Government.

Regional Australia Summit Steering Committee (2000) *Regional Australia Summit Final Report December 2000*, Canberra: Department of Transport and Regional Services.

Stewart-Weeks, M. (2000) 'Trick or Treat? Social capital, leadership and the new public policy', in Winter, I. (ed.) *Social Capital and Public Policy in Australia*, Melbourne: Australian Institute of Family Studies.

Chapter 6

Faith Based Service Programs in the Public Sector: Illustrations from Virginia of Faith in Action

Randall Prior

Introduction

Each workday and most evenings at churches and synagogues around the United States the community gathers. People gather for meetings of Alcoholics Anonymous and Al-Anon. They gather for English as a Second Language classes, for support groups, tutoring for SAT tests, and for parenting or job search seminars. Parents bring their young children to preschools and elementary school programs. What is significant about these offerings is that in many cases the participants have no other association with the host congregation. They come because something is offered that meets their need in a setting they trust to be safe, and concerned. They come because there remains some awareness in our culture that faith groups have a commitment to build community, address the needs of individuals, and value and respect the dignity of each person.

The prophets in the Hebrew scriptures declare the necessity of concern for the sojourner and stranger in the community. They pronounce warnings for those who rest comfortably in self-absorbed affluence and piety and forget the needs of those around them. Faith communities are called by God to seek justice and mercy for all in the larger community in which they exist.

The New Testament parable of the Good Samaritan (Luke 10:25-37) serves as testimony that one's neighbors are not just the people you know and like, but in fact are all others regardless of who they are or where they have come from. The model taken from that parable and from the events of Jesus' ministry is for a community without boundaries in which poor and dispossessed are served and where there are no outcasts. These ideals are admittedly realized imperfectly in many congregations where cliquishness, prejudice, and even xenophobia may sometimes abound. Nonetheless, the mandate of faith communities is to work toward a just society in which mutual respect, understanding, and compassion are hallmarks.

Religious groups have answered this call over the centuries in the founding of hospitals, orphanages, homes for the disabled, elderly, or mentally ill, soup kitchens, and hospice ministries. Societal growth and change has often meant more people are left behind in an increasingly affluent culture. The dependence on government and other non-religious resources has been essential, but even public resources prove inadequate. The monumental needs and complexity of issues in this age have long since overrun the capacity of any single sector to adequately address them. The need for partnerships of cooperation and collaborative effort to mitigate human suffering is clear.

This occurs, as we shall see, in partnerships of religious organizations or faith groups with each other. It also occurs in partnerships of faith groups with business and government. These combined resources can most effectively address the needs of communities that are increasingly diverse in terms of culture, religion, outlook, and interests. This chapter will provide concrete examples of typical ways faith based social service programs impact the public sector and in so doing answer the call to service inherent in their religious traditions. The illustrations focus on the church sponsored school, direct service organizations, Habitat for Humanity, the religious community in the public policy debate, and finally, faith based organizations and the global community.

The church sponsored school

Church schools are often viewed as institutions primarily dedicated to the propagation of a particular faith tradition or even particular perspectives within a faith tradition. This is one function they serve. However, in a broader sense many church schools seek to provide environments in which children from varied backgrounds and needs can learn and have the experience of relating to a faith based community of peers. Faith based schools often approach their task with a particular dedication to individual and special needs.

St. Andrew's Episcopal Preschool, Burke, Virginia exists in a relatively affluent Washington, D.C. suburb which in recent decades has become increasingly diverse in terms of race and culture. The school has placed a conscious value on diversity in all aspects of its program. Scholarship funds are explicitly directed toward families not able to afford preschool education for their children and tuition assistance is provided to meet short term needs. Community groups and government agencies are often solicited in an effort to find children eligible for these funds. It is common to see "shadow teachers" with autistic children or children with physical or emotional disabilities or developmental delays as part of classes. The school has served children with Downs syndrome, brittle bone disease, leukemia, severe heart defects, delays in speech, language and fine motor coordination.

The school director and staff work closely with the local school district's early intervention program to implement appropriate Individual Educational Plans for students with special needs, or to give professional input to assist with adequate future placement for them. Often diagnosis and treatment occur as a result of the school staff urging parents to seek testing and evaluation. Parents are supported by teachers as they come to terms with their child's difficulty. All children in the school benefit from an environment that is inclusive and diverse in terms of different ability peers. School classrooms are filled with children who reflect a broad cross section of cultures, races, and religions, though the school consciously operates out of a non-dogmatic Christian perspective.

The school seeks to assist parents in learning about the development of young children. Regular parenting workshops and a library on parenting assist families of students. Within the church program the presence of parenting support groups, ESL (English as a Second Language) or other vital family services are a further avenue for assistance.

The staff and director provide referrals to individual families regarding the acquisition of health insurance, for a used vehicle if the family is without transportation, or to family counseling in times of crisis. Annually the school participates in the community observance of the Week of the Young Child, an observance sponsored by the National Association for the Education of Young Children (NAEYC). This event highlights work done in preschool education. The director represents the school in a coalition of preschools that advocate for the importance of early childhood education and the school is accredited by the NAEYC and so adheres to the highest standards of preschool operation.

The school has its own outreach or community building projects. A collection of hats and mittens is taken at Christmas time for needy children and a collection of children's books that are donated to a shelter for abused women and children. Classes adopt local nursing homes and visit regularly to bring joy to residents and to learn from another generation.

Only 10 per cent of the student population are from families that are members of the sponsoring congregation. Community input to the school's operation is gained through a board representative that is elected from the community at large. The parish sees the school as an extension of its outreach to the community and in no way relies on it as a for-profit venture. In fact, it is not. However, what is accomplished in the lives of young children, for struggling families, and in the future of the community at large is a rich return indeed.

Direct service organizations

One of the most familiar interventions of congregations in the life of communities is the direct service organization. This can be a program sponsored by individual congregations or by coalitions of congregations from different faith traditions. This type of program is typically the soup kitchen, the homeless shelter, or meals on wheels for the sick and shut in. These programs often include assistance with clothing and furniture, emergency rent subsidies, or low cost automobiles to meet transportation needs. In many communities food and clothing are warehoused in a particular church facility or in a building jointly owned by a church coalition. Such facilities serve families on the basis of severity of need.

In other instances such organizations might offer job counseling, health clinic services, or transitional housing for families that are struggling to get back on their feet after a major loss or trauma. In one particular program, Fairfax Area Christian Emergency Transitional Services (FACETS), Fairfax County, Virginia, a coalition of congregations work together to feed more than 150 people every night of the year. They provide tutoring and mentoring services, school supplies for needy children, and job counseling and home management training for youth and adults. This program works closely with the county social services department and facilitates the link between the church and government by targeting resources to meet needs most effectively. This type of program is manna in the wilderness, bread for the journey, in lives beset with challenges and troubles. It is a clear example of faith commitment motivating meaningful intervention in the lives of others.

Habitat for Humanity

One of the best-known faith based groups is Habitat for Humanity International (HFHI). Founded a quarter of a century ago by Millard Fuller, an entrepreneurial member of a Christian community in Americus, Georgia, Habitat has grown into an organization that builds "simple, decent, affordable, housing" in seventy countries around the globe. By 2001 Habitat had finished its 100,000th house world wide and set its sights on completing another 100,000 in 20 years. The goal of the organization is to eradicate inadequate housing worldwide. It operates through a global network of local affiliates and regions, which generate funds and volunteer efforts to erect homes in their area.

Habitat provides houses for people on the basis of need. Individuals and family are expected to partner with a local Habitat affiliate as part of the process which requires that they perform 300-500 hours of "sweat equity" work on another person's Habitat home and their own. Homes are sold to the partner families on a no interest loan basis usually at prices at or below the cost of construction. A one percent down payment is required. In this way families living on the margins of society with incomes 1/4 to 1/2 of the median income for a given area are able to realize the dream of home ownership.

From Habitat's perspective this ministry is about more than building houses, it is about building a future for families and children. Increased family stability and economic opportunity comes with homeownership. In addition, self esteem and community consciousness are raised in new homeowner families and are the real long term benefit of this work which, in turn, affects the larger community as well. Habitat for Humanity is the catalyst for meaningful, constructive interaction of congregations, corporations, community groups, and local governments. Funding for house construction is provided by a sponsoring entity, often a corporation, community service club, or an interdenominational cluster of congregations. On occasions affluent individual families have contributed the total cost of house sponsorship. House construction is done by volunteers many of whom come from the sponsoring organizations. Sponsors and future homeowners work side by side as partners and colleagues in the adventure of building, thus uniting various parts of a community in meeting housing needs and in building the community.

Local governments which have a vested interest in affordable housing frequently facilitate or contribute to land acquisition and design review requirements. As a faith based organization, Habitat has engaged in partnership with the federal government in recent years through the Self-Help Homeownership Opportunity Program (SHOP) and the Capacity Building for Habitat for Humanity program administered by the Department of Housing and Urban Development. The HFHI statement of faith based legislation defines the use of these funds as follows:

> These funds have provided seed money to help local Habitat affiliates "set the stage" for house building through land acquisition, infrastructure and increased capacity. Moreover, this government assistance has been leveraged by Habitat into millions of dollars of private support for home construction in many of the 1,600 communities in which Habitat is building in the USA. This country has the resources to alleviate suffering of millions of Americans living in overcrowded, substandard and unaffordable conditions. But no one organization by itself can end poverty housing. The solution lies in collaboration, with all sectors of the society working together.

Local governments serve as a channel for loans or grants through HUD or other local housing funding programs to assist the project funding and overhead of the local Habitat affiliate. Most affiliates see their relationship with homeowner families as ongoing. Financial planning, home repair and management assistance is provided to families in advance of house completion. Assistance with taxes, wills, and other legal matters is also given, if needed. Community service groups, financial planners, attorneys and others with needed skills are solicited as volunteers in these efforts. Support continues for families following settlement, as Habitat often continues as mortgage holder with a vested interest in not only the house, but also the family's success in their new home and new life.

The power of a community coming together as non-profit sector, government, and business is no better demonstrated than in a program like this. Each sector gives up some control over the process but each also gains in being drawn more fully into a community of relationships and cooperation that accomplishes a purpose none could have accomplished alone. Each sector also gains in furthering its own mission. Congregations have opportunity to carry out their mandate to serve those in need. Corporations demonstrate concern for their community, achieve beneficial public relations, and create opportunities for corporate team building as their departments work together as volunteers on home sites. Construction related businesses often gain visibility as the donors of building materials and services. Government housing agencies have the benefit of private funding assistance and energetic community support in facing the pressing need for affordable housing units in their area.

One such affiliate, Habitat For Humanity of Northern Virginia, was founded in 1990 to serve the urbanized counties of Arlington and Fairfax and the City of

Alexandria. In that time 29 homes have been built serving nearly 120 people. Developments of three to eight units as well as several individual homes have been constructed in an area which is heavily developed and in which land and housing costs are well above the national average. The affiliate's most recent and largest development, currently under construction, will provide townhomes for eighteen families. The affiliate's strategic plan projects building over 20 houses a year by 2005 with the support of a strong and diverse religious community, strong ties to the local housing authorities, and generous corporate participation.

One typical example of the way that Habitat serves people in need is embodied by the case of J.P. and C.K. They came to the United States as political refugees from the Congo in the early 1990s. They had been active in the pro-democracy movement in the Congo and fled their home after being beaten and threatened. A condition of their political asylum was that they were unable to accept any welfare or state assistance. Although J.P. was an accountant in the Congo he was initially only able to find work as a janitor at a local Baptist Church, which was one of Habitat's partners. They and their four young children became members of a Presbyterian Church in Arlington, another of Habitat's partner churches. Subsequently the couple became involved with Habitat as volunteers on a work site. The family applied for a Habitat home in 1998 and moved into their new home in Arlington in October 2000. This family has four children ages 6-14 and hope eventually to bring the two children of J.P.'s brother to this country. J.P.'s brother was assassinated in 1992 for his political activities in the Congo. A consulting firm in Arlington,Virginia, was the Habitat sponsor for J.P. and C.K.'s house and was able to find J.P. a position in their accounting department where he continues to be employed today. C.K., his wife, has most recently worked at an Alexandria bank. Their children all attend Arlington Public Schools in their neighborhood and they continue to attend Trinity.

The religious community in the public policy debate

Religious groups in some states form organizations to address issues of concern and to provide information, public education, and advocacy on important issues before state and local governing bodies. Once such organization is the Virginia Interfaith Center for Public Policy (VICPP). This body which operates in cooperation with the Virginia Council of Churches was founded in 1982 after many informal partnerships among church groups. It identifies itself as 'the only non-profit, statewide organization advocating Judeo, Christian, and Islamic values'. It seeks to improve the laws and programs of the state guided by 'a vision of a more just and compassionate Commonwealth'.

VICPP cites among its recent accomplishments tax relief for the working poor, increase in public assistance benefits and funding for daycare, indefinite time limits for admission of DNA evidence in criminal cases, and increased funding to assist the homeless. In addition to these issues the organization's current (2002) priorities include:

1. reduction of the number of families among the working poor or living in poverty,
2. abolition of the death penalty,
3. increased services for at-risk children and youth,
4. extension of adequate physical and mental health care to all Virginians,
5. increased availability of low cost housing.

This organization provides an ongoing presence in the Virginia State capital during sessions of the state legislature. It presents the perspective of diverse faith communities to individual legislators and committees. It also provides (through newsletters and its web site) up to date information on a number of bills that it follows through the session. Information is provided to constituent member faith groups about VICPP concerns regarding each bill and contact information about assembly committee membership for those who wish to express opinions to their representatives.

The Center recognizes the delicate balance of First Amendment liberties in its work and the standard of institutional separation. But, in the words of VICPP executive director, Fletcher Lowe, 'on matters of social policy where the lives of those for whom God has a bias (as described in each of the sacred scriptures of the Center), we move to find that common ground of cooperation … . We always need to remember our prophetic role of calling the State to accountability, which may place us against the decrees of the state. But we need not be reluctant, for even though we come at the (church-state) intersection from different perspectives, we share - faith community and legislators - the common ground of those most vulnerable in our Commonwealth of Virginia'.

Faith based organizations and the global community

Religious communities are global in nature. Christian churches and missions, Jewish communities, and increasingly Muslim, Buddhist and other congregations exist in multicultural communities across the United States and much of the western world. International partnerships between people of the same, or even different faiths, are increasingly common. In most cases there is a reciprocal benefit realized in these global partnerships. For example, in the Christian community it is common for western technology, financial resources, and expertise to be sent to churches in developing countries. This often provides programs which enhance living standards, supply medical care, and fund education and other public service programs. Less obvious, but of equal and sometimes greater long term value, is the way these partnerships create interactions between cultures that are a direct benefit to the developed world.

Christians living in societies in which they are a distinct minority, and often in places where life is difficult, have a much different outlook on their faith than those in traditionally Judeo-Christian nations. Often the well-developed, multicultural insights and survival skills of people of faith in other lands provide an important experience for westerners faced with an ever more diverse culture to which they must adjust.

In a remote area of western Tanzania, East Africa, a typical partnership has existed between the Diocese of Western Tanganyika (DWT) of the Anglican Communion and several parishes of the Episcopal Diocese of Virginia in the United States. This African diocese led by the Rt. Rev. Gerard Mpango is a model of effective sensitive ministry devoted to making a difference in the lives of the people it serves not only in the church but in the villages, towns, and rural areas of the region. Bishop Mpango who grew up in a rural Tanzanian village has become a savvy and sophisticated agent of community development and church growth in this area.

Since the early 1970s churches in the northern Virginia area have worked together to supply funding for programs in the DWT. Such programs have included support for the Bible School in Kasulu which trains lay catechists who go into the remote villages to teach and preach. It has provided funding to bring clergy from that diocese to the United States or England for further theological education.

A short wave radio system was provided to better link the remote missionary hospitals with other mission posts. Some of these outposts are staffed by the German Free Church in cooperation with Anglicans, Baptists and others in the area. These are truly remarkable examples of international and interdenominational effort. The radios also serve to coordinate the efforts of another mission service called MAF (Mission Aviation Fellowship) a group that flies small planes around East Africa to provide transportation both for residents and visitors, and also medical evacuation and transportation assistance for the sick or injured.

Public service projects sponsored by the DWT which serve the population at large have been underwritten by the American partners. Among these are funds for purchase and husbandry of poultry, dairy cows, and pigs to increase the protein in the local diet and provide opportunity for individual entrepreneurship and jobs in the local area. A large lorry was purchased for the diocese to enable the transport of building and food supplies, cattle, generators and other goods over very rough, rut laden roads in the area. Habitat for Humanity International in the Kasulu area of the DWT has been heavily funded from the United States. Numerous sturdy houses built of concrete and brick now replace the very transient stick, mud, and thatched roof construction common to the area.

Another organization related to international bridge building and resource development is Five Talents International. This effort uses funds from the developed world to establish microeconomic development banks in numerous places throughout the developing world most often under the sponsorship of a local religious leader. Clusters of people are then granted small loans to establish cottage industry businesses. The resulting income provides for them and their families.

Loans mostly go to acquire equipment needed to utilize skills already present or easily developed. Such equipment might include sewing machines, woodworking tools, agricultural or animal husbandry supplies.

In this and other ways relationships across the face of the earth are developed and can grow. The bonds of understanding and support replace the alienation and misunderstanding that comes from unfamiliarity with each other. Our encounter with the global community is a chastening and enriching occasion of growth. We learn from them what it means to be more effective in the call of our religious life and ministry. We learn from them what it means to do much more with much less. Finally, we learn that family, community, and growth can happen even more effectively in environments in which the challenges are greater and the needs much more pressing.

Conclusion

The illustrations provided above are typical of the ways that faith based service programs relate to a larger community of business and government in carrying out common purposes of community involvement and development. Religious institutions share a common stewardship of life and resources with the other sectors in society and while the ultimate purpose and mission of each sector may vary considerably, they are united in a common concern to bring their resources to bear upon the problems of society and, through the opportunities inherent in working together, to make a better life for all. This effort, surely, must continue to represent a prominent response to the broader challenges of participatory governance!

Section Three: Operational Dimensions of Participatory Governance

Chapter 7

Regional Transportation Strategies in the Washington, D.C. Area: When Will They be Ready to Collaborate?

W. Robert Lovan

Background

This chapter focuses on a public decision-making process involved with planning and implementing a regional metropolitan transportation strategy. The process is made difficult by other public policy issues and further compounded by inter-jurisdictional interests.

Since the 1920s, community leaders in the Washington metropolitan region, (Northern Virginia, Central Maryland and the District of Columbia) have recognized that interconnected economic factors determine the quality of life for their jurisdictions (Figure 7.1). For more than a decade, however, the single regional public issue has been the coordination of transportation planning. It has not focused on meeting the needs of a regional economy. Instead, the debate is consumed by transportation issues and the impact of the attendant transportation infrastructure on the quality-of-life for the metropolitan region.

Two important public policy questions are driving this regional, transportation public decision-making debate – land-use issues and environmental issues. The land-use question concerns how various modes of transportation,[1] and the attendant infrastructure, will influence development densities and commuter patterns. Closely tied to this land-use issue is the quality of the environment – particularly air and water quality. Population growth and the burning of carbon fuels by the internal combustion engine (automobiles, buses and trucks) are threatening the regional air quality to the degree that the Federal Government has moved to sanction the regional governments.[2] In addition, urban domestic chemical usage and other residual pollutants, a by-product of the urban sprawl (individual usage in home and lawn applications), threatens the quality of the area watershed.[3]

Over and above the strains on the public decision-making process to resolve the transportation planning issues (compounded by land-use, air and water quality issues and to a lesser degree economic development strategies), the debate is further fractured by the interests of the many governmental entities. They include the States of Maryland and Virginia, Washington, DC as a semi-autonomous region

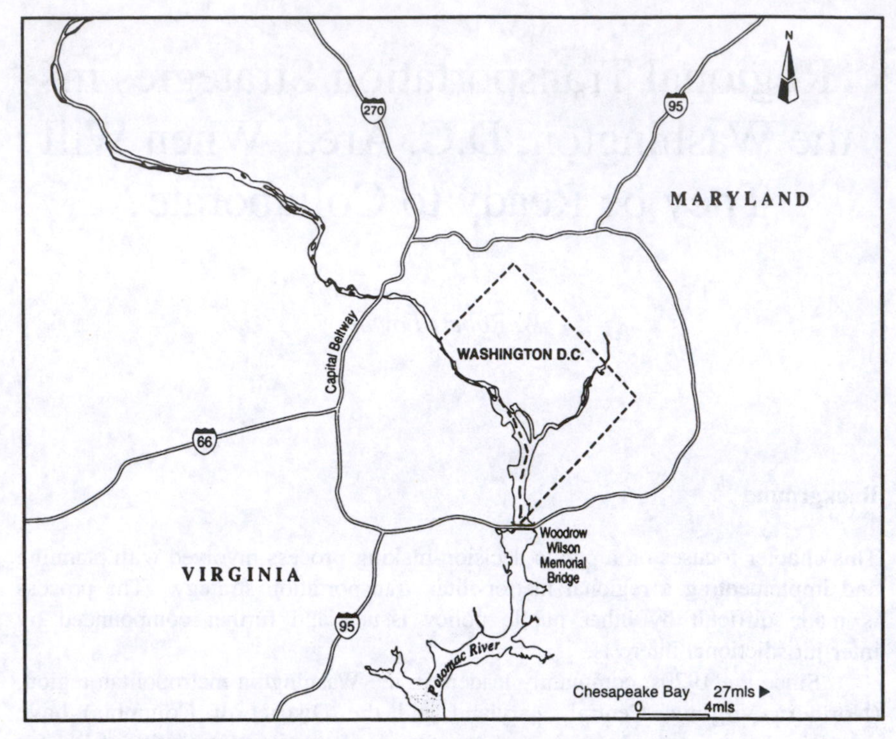

Figure 7.1 The Regional Context of Washington, D.C.

dominated by the Federal Government, seven counties and 25 local governments. Each of these 35 governments holds a form of veto power within the public policy decision-making process.[4]

This attempt to coordinate inter-governmental, public policy decision-making for the region is not a recent issue. The first regional meetings were held in 1922. In 1957, responding to inter-jurisdictional decision-making needs, an intergovernmental organization, the Metropolitan Washington Council of Governments (WashCOG), was formed.[5] Regional transportation planning was one of the first items on the new metropolitan council's agenda. Transportation planning was assigned to the National Capital Region Transportation Policy Planning Board (TPB). Beginning in the 1960s, for example, in order for the metropolitan region to obtain Federal assistance for any desired transportation project, the regional entities were required to coordinate all of their individual transportation projects.

This Federal mandate not only included improved intergovernmental coordination, but also "required citizen involvement and the directive to eliminate damage to the environment" (Weiner, 1997). In addition to the Federal mandate, there are also state-level intergovernmental planning bodies in Maryland (the National Capital Area Parks and Planning Commission) and in Virginia (the Northern Virginia Regional Commission).

During this same period, non-governmental private local, state-level and regional organizations were formed to advise and advocate a particular transportation policy. One of the first was the Greater Washington Board of Trade representing the business sector. Today, there are dozens of prominent inter-regional advocacy organizations representing the interests of the three sectors of the civil society (government, associational and business).[6] Each group campaigns for its own preferred transportation, land-use/environmental protection and/or economic development program. In addition, three other influential inter-regional non-governmental organizations are from the associational sector: the Piedmont Environmental Council, the Chesapeake Bay Foundation and the Coalition for Smart Growth (ICAR Transportation Study Team, 2000).

Yet, at the end of 2002, with all of this structure for comments by citizens and sectors to the public decision-making process, the three principal jurisdictions (Maryland, Virginia and the District of Columbia) remain more divided than ever over what public policy strategies are to be devolved than they are united over what transportation projects to implement. Since the 1960s, the single regional project to be implemented is the current construction to replace the deteriorating Woodrow Wilson Memorial Bridge (built in the 1950s), over the Potomac river at Alexandria, Virginia. Also, during the 1960s, the beltway around the District was constructed (Figure 7.1) and the decades long work on the metropolitan region Metro rail/bus system was begun.[7]

These 1960 infrastructure projects were meant to aid the flow of the hundreds of thousands who commuted to work daily from Maryland and Virginia into the District of Columbia. Today, however, more jobs now exist outside of the District than inside. The 2000 U.S. Census also showed that the Virginia and Maryland jurisdictions have five times more residents than the District (WashCOG, 2001). The center no longer commands the whole. Frustration with regional congestion has risen to the point that the transportation problems are now interconnected with land-use and the environment issues, all of which have now become difficult political issues. This three-part debate[8] (transportation, land-use and the environment) is taking place across and within many arenas in the region: (1) within and among the three sectors (government, associational and business), (2) within and among the local and the State levels of government, as well as (3) within and between Maryland and Virginia, and to a lesser degree, the District of Columbia.

Understanding the conflict

Since 1988, the Institute for Conflict Analysis and Resolution (ICAR) at George Mason University[9] has studied and intervened on issues of regional cooperation and coordination in the metropolitan region. On the transportation planning issue, during the 1999-2000 academic year, graduate students[10] conducted structured, open-ended interviews of dozens of recognized public leaders[11] in Maryland, Virginia and the District of Columbia. These respondents were drawn from the government, associational and the business sectors. They were asked a series of questions as to why regional transportation planning and cooperation was so difficult. The interviews began with transportation policy questions and expanded to the connected issues of land-use, the environment and economic development.

Information from these interviews was summarized, patterned responses identified, and then from this analysis eight broad areas of concern were identified. The interview data concentrated on what was believed to be the *characteristic difficulties* with regional transportation planning (ICAR Transportation Study Team, 2000):

- *Multi-jurisdictional character* of the Washington metropolitan region, recognizing the region includes numerous local jurisdictions, the District of Columbia government and two state governments.
- *Incomplete and ineffective citizen review* as a concern of the elected officials. Two concerns were expressed: Do the procedures required to be followed in fact represent the interests of the citizens? Do the participating individuals and the groups (representing sectoral interests) constitute more than a 'superficial' citizen consultation process?
- *Divided authority* spreads out (thins) the policy making process. Planning, decision-making and implementation among the numerous local governments, sub-regional entities, and state and multi-state/inter-regional authorities limit collaborative decision-making.
- *Mandated processes* as outlined by Federal and State governments. There are too many points in the process where a previous decision can be challenged and is challenged.
- *Limited* [public] *funding* represents insufficient public fiscal resources to implement all of the transportation projects nominated by the many governmental entities. Consequently this directly causes competition and confrontational interactions among the sectoral participants.
- *Already limited communication* is compounded by conflicting views. Incomplete communication often results from differing logic and perceptions of technical terms and procedures employed by politicians, engineers, advocacy groups and citizens when communicating with each other. The challenge is understanding is, how different interests react differently to seemingly similar information.
- *Numerous stakeholders* representing: (1) citizens and elected officials, who must articulate transportation needs; (2) State and local elected officials who must prioritize projects; (3) transportation planners who must prepare the technical plans; (4) regional boards that must assimilate local and sub-regional initiatives into a comprehensive regional vision; (5) Federal and State officials who allocate fiscal resources; while, (6) consultations with civic, environmental, business and other special interest are required at every step in the project development and funding process.

- *Circular decision-making process.* The required process appears to many to be a never-ending cycle rather than following a set of specific, discrete and sequential decision-making steps which will lead to an agreed end result or product.

Following the compilation of the list of these eight areas of characteristic difficulties the ICAR transportation study team met with the Council of Governments TPB. The team asked the TPB to concentrate on those factors on which the Board believed they could take action. A critical question comprised the areas in which the Board believed that collaborative action among the three sectors could be undertaken. Table 7.1 identifies a number of the contributing and inhibiting factors for taking collaborative action on the characteristic difficulties. (ICAR Transportation Study Team, 2000).

Table 7.1 Factors Affecting Regional Transportation Planning

Factors which *Contribute* to the Planning Process	Factors which *Inhibit* the Planning Process
• The ability of the TPB to reach consensus and make decisions despite enormous complexity.	• Insufficient funding.
• The requirement to conform to mandated procedures provides momentum to achieve project implementation.	• Different governmental structures and decision-making among jurisdictions within the metropolitan area.
• Each governmental jurisdiction embraces its own philosophies and objectives and generally meets the goals it establishes for itself.	• Parochialism between elected officials and constituencies.
• Lack of public support to take the measures necessary to bring about real changes.	• Differing values espoused by jurisdictions within the metropolitan area.

It is interesting to note that during the discussion one Board member said that the process employed by TPB seemed to be functioning as designated, but that the process increasingly seemed to result in unsatisfactory products. Several Board members indicated that TPB's work pattern does not allow for deeper policy discussion of some important issues, and in fact erects barriers to the requirement in reaching a regional agreement.

At the end of the meeting, the ICAR transportation study team raised the question as to how the TPB might use its authority for dealing with the identified characteristic difficulties. Three approaches for sorting out actions were identified: (1) where TPB has the authority, but because they are unaware of the situation they take no action; (2) where TPB has the authority, but decides to not take action; and (3) where TPB does not have the authority and, therefore, should not take any action.

The ICAR transportation study team suggested a strategy for TPB to use in dealing with the characteristic difficulties of competing issues and interests within the transportation debate. A process was proposed which would allow for a collaborative discussion concerning transportation planning and the associated issues. Essential participants from each of the three sectors would participate in the discussion. The objectives would be:

- to support networks for informal communication and relationship building in a neutral and relaxed atmosphere;
- to identify the obstacles and the opportunities for effective multi-sectoral decision-making; and,
- to generate options for overcoming the obstacles and capitalize upon the opportunities.

An inter-sector dialogue

During the 2000-2001 academic year, ICAR explored the potential for the presidents of the two major state universities in the metropolitan region, George Mason University (Virginia) and the University of Maryland, to offer their institutions as neutral sites for a series of dialogues among regional leaders. State partisan politics proved too much of a risk for the universities, so invitations were never extended. In addition, coming state elections in Virginia suggested that it might be difficult to get public leaders to apportion their scarce time to attend even a few brief dialogue meetings.

It was decided, in order to overcome these limitations, it would be necessary for ICAR to offer to conduct a virtual conversation among a target number of public leaders. The discussions would be made possible by the use of a three-round Delphi process[12] (Adler and Ziqlro, 1996).

The initial participants in the Delphi were selected from the list of community leaders who were interviewed by the ICAR transportation study team. They were instructed that the Delphi process offered them the opportunity to have a dialogue with other leaders from across the three sectors but without having to attend a meeting. The participants were asked to represent their own organization, not their personal views. Eventually more than 90 leaders from the government, business and associational sectors in Maryland, Virginia and the District of Columbia participated in the discussion.

Delphi statements in the first round were constructed from the earlier ICAR transportation study team interview data. These statements were focused on the problems of coordinating transportation planning and directed the respondents to consider goals for a coordinated transportation planning effort. Participants were asked to agree or to disagree with the statements and to name additional factors that make it hard, or easy, for jurisdictions to cooperate. They were asked to describe their perceptions of leadership roles and responsibilities, as well as the organization rules and resources of the typical group from each of the three sectors (government, associational and business). They were asked to describe what could be done now, and what would have to be changed to facilitate collaboration. Finally, they were asked to name people and organizations they thought should be involved in the Delphi dialogue.

In the second Delphi round, the lightly summarized responses[13] from each of the nine sections in the first round were fed back to the participants. They were then asked to rank the various responses in terms of their agreement with the particular statement, to indicate the importance they gave it, and if they wanted to challenge any information from their fellow participants. This structure was intended to simulate addressing each other in an actual meeting.

In the third Delphi round, participants were asked to affirm or challenge the consolidated responses[14] from the previous round. They were also asked to indicate if their organization would participate in a regional consensus-building effort. This would be based on the principles and issues identified in all three-rounds.

The level of participation by the various groupings in all three rounds was steady: (1) from the three sectors (government, associational and business); (2) from the various interest (business, community, civic/citizen, environmental and the levels of government – local, State and Federal); as well as, (3) from the Maryland and Virginia geographical areas, but not the District of Columbia. Transportation planners in all three jurisdictions were the most active participants. Numbers of survey respondents increased as participants recommended additional respondents. The actual percentage of responses per round and per grouping remained high throughout all three rounds. In the final survey, the number of contributing organizations included eight from the District of Columbia, 29 from Maryland and 37 from Virginia. Fifteen regional or multi-state groups were also represented for a total of 89 entities.[15]

Delphi discussions

In the third and final round of the Delphi, leaders were also asked to affirm or challenge the statements built from the dialogue in the earlier two rounds. The following indicate the "most agreement":

- *Many dedicated individuals and organizations are employing substantial energy to make transportation in the metropolitan region safer, effective and compatible with the environment. Yet, the organizations involved in effecting that change have trouble collaborating on what is to be accomplished. Basic differences, and therefore political conflicts, exist over balancing economic development, land-use, the environment with the desired mode(s) of transportation* (93 percent of the respondents agreed, 7 percent had no response).

- *Collaboration*[16] *to build and sustain the political will to commit to a regional solution should be a part of the process to improve transportation planning and coordination* (93 percent of the respondents agreed, 7 percent disagreed).
- *A strong focus is required on providing access to jobs, services, shopping and recreation – not just highway projects* (89 percent of the respondents agreed, 11 percent disagreed).
- *An equal emphasis on and the importance of both land-use planning and transportation planning* (83 percent of the respondents agreed, 12 percent disagreed and 5 percent had no opinion).
- *More effective citizen involvement* (83 percent of the respondents agreed, 14 percent disagreed and 3 percent had no opinion).

In each round leaders were asked to list, explain and prioritize up to three critical factors which make regional cooperation difficult. In the last round, rankings from the previous rounds were listed and the participants were asked to again rank the final list. The three top factors making regional cooperation difficult were:

- *The multi-jurisdictional character of the region* (86 percent of the respondents agreed).
- *Divided authority* (69 percent of the respondents agreed).
- *Limited funding* (69 percent of the respondents).

The participants were also asked if their organization would be interested in participating in a collaborative process with the purpose being to work on implementing the goal drafted by the participants in the three rounds. Seventy-nine percent agreed with the goal and were willing to be involved (Table 7.2) When the public leaders were asked who should sponsor and implement such a consensus-building process, participants had diverse responses (Table 7.3).

Table 7.2 Goal of a Collaborative Process

Goal: To engage public and private decision-makers in a quiet, inter-governmental and inter-jurisdictional process to achieve a broad consensus for effective and comprehensive regional transportation planning, implementation and management.

Respondent agreement/disagreement with this goal:
79 percent agreed with this goal
17 percent express reservations
3 percent had no opinion

Table 7.3 Who Should Lead a Consensus-Building Process?

	Lead	Sponsor
State Government	76 percent	69 percent
Local Government	69 percent	69 percent
Transportation Agencies	55 percent	66 percent
Federal Government Agencies	34 percent	52 percent
Business Organizations	24 percent	45 percent
General Citizen Groups	14 percent	31 percent
Environmental Interests	10 percent	28 percent

There was significant agreement that State governments should both lead (76 percent) and sponsor (69 percent) an inter-governmental process, but that interest organizations should not. Most agreed that all of the sectors and interest groups should participate.

Finally, the respondents were asked, if a dialogue were to be convened to work toward their agreed goal for resolving the issues identified in the Delphi, and if led and sponsored according to the preferences they had indicated, would their organization participate?

- 97 percent said ‘yes’ (69 percent were strongly supportive, and 28 percent were mildly supportive).
- 3 percent were mildly opposed.
- No one expressed a no opinion about participating.

The challenges: coming into a sharper focus

Neither the structured interviews nor the three-round Delphi was an attempt to determine the beliefs and attitudes of the general public concerning the difficulties in planning and implementing transportation solutions for the metropolitan region. The number of participants was too small to draw statistically valid conclusions about the views of citizens in the region. The findings appeared to be self-evident: 'what is already known!'. The findings, however, appear to be validated as the public debate continues, as citizen groups speak out and as various public events unfold. There is much to be learned/confirmed from the responses of the leaders from the three sectors. In fact, they did exhibit considerable agreement on the following challenges:

- Cooperation at the top is needed to stimulate collaboration among the other players. There is hope that new administrations[17] in Virginia, Maryland and the District of Columbia will work together better than these jurisdictions have done in the past.
- Inter-jurisdictional processes should seriously examine both the technical and the political structures and barriers to cooperative efforts.
- Transportation planning should be integrated in a purposeful manner with land-use, environmental and economic planning. Efforts need to be implemented which will go beyond limited actions and projects. Such an effort will require the demonstration of real communication and understanding among the three sectors.
- Business, civic and environmental groups should equally participate, but none can lead the consensus-building process. This process has to be managed and facilitated by a true neutral.
- While funding is viewed as a major complicating factor, it was not the most often cited factor. Most leaders indicated that the lack of funding was related to the lack of public confidence in the planning and implementation mechanisms. They also mentioned the process was incapable of dealing with land-use and environmental issues.
- Even with an expressed willingness to work with the other groups across the sectors, there is considerable suspicion about the agendas of the other sectoral groups. There is a very real lack of trust (understanding) by interest groups from one sector toward the groups from the other sectors.

Update and conclusions

For many long-term participants in the struggle to provide public transportation solutions for the region the results of these consultations (the interviews and the Delphi) are not surprising. Results strongly indicate that these active organizational leaders from all three sectors want their elected representatives to lead, to be more active in helping to form a consensus, and to do more than just respond to others.

They want to support their governors, mayors, county executives and legislators. It is recognized that there are both differing leadership capacities (roles and responsibilities), as well as differing organizational capacities (rules and resources) across the three sectors and within the three jurisdictions. There is an expectation that, through collaboration, shared strengths can overcome any perceived weaknesses. In fact, some from the non-governmental, business and associational groups, have become so frustrated by inaction that they have tried to lead themselves while at the same time acknowledging this is a second best approach.

Based on the assumption that funding would break the logjam, the Commonwealth of Virginia legislature gave the Northern Virginia leaders authority to schedule a referendum on the creation of a new taxing district to fund transportation projects for that portion of the region.[18] It was believed that a new taxing district (an additional half cent per dollar sales tax) would go a long way in solving the Northern Virginia problem. This approach, however, was not what the ICAR consultations indicated. In fact, funding was not seen as the principal issue, albeit a significant one. Neither the potential taxing mechanism, nor the funding it would provide, would offer much support for dealing with the other policy issues (land-use, environment and economic development) that make transportation planning such a difficult public policy issue. Subsequently, by a substantial 55 percent to 45 percent margin, Northern Virginia voters decisively defeated the proposed new taxing district. The lure of designating funding for transportation projects was not enough to overcome the other issues.

A major factor in the equation remains the cleanup of the metropolitan regional environment. The Federal government has weighted the environmental portion of the dialogue by reclassifying the region as 'being severely below Federal air quality standards'. The action could result in the withholding of Federal funds for transportation projects.

In the near term cooperation without the new regional taxing authority[19] may be more possible than ever. Virginia's new governor appears to be more sympathetic to the broader set of cross-sectoral and cross-State issues than his predecessor. Also, Maryland's new governor has indicated a constructive interest in dealing with the complex and competing regional transportation issues. And, the District's re-elected mayor has always supported finding regional transportation solutions. As the ICAR findings indicate, these public leaders are seen as critical in providing the motivation for building collaboration in evolving strategies to resolve the transportation conflicts.

Building trust appears to be the biggest need and at the same time the biggest challenge. Honest communication in a secure and neutral setting is critical for building an understanding of the basic issues across the three sectors, even if it is not possible to build collaboration between the many groups and across the sectors. The information from the ICAR consultations among the sectoral leaders should reassure top elected policy-makers and opinion leaders of one major finding. If senior policy-makers indicate that they are serious about building collaboration across the jurisdictions and among sectors to resolve regional transportation congestion, they will likely find broad political support for their efforts.

References

Adler, M. and Ziqlro, E. (eds) (1996) *Gazing into the oracle: the Delphi method and its application on social policy and public health*, London: Jessica Kingsley Publishers.

Hughes, J. Knox, C. Murray, M. and Greer, J. (1998) *Partnership governance in Northern Ireland: the path to peace*, Dublin: Oak Tree Press.

ICAR Transportation Study Team (2000) *Transportation planning and coordination in the metropolitan Washington, D.C. area: what makes it so hard?*, Institute for Conflict Analysis and Resolution, George Mason University, Virginia.

Katz, B. (ed) (1996) *Reflections on regionalism*, Washington, D.C.: Brookings Institute.

Mattessich. P. and Monsey, B. (1992) *Collaboration: what makes it work*, St. Paul, MN: Amherst H. Wilder Foundation.

WashCOG: Metropolitan Washington Council of Governments (2002) WWW.WashCOG.org. Washington, D.C.

WashCOG: Metropolitan Washington Council of Governments (2001) *Our changing region*, (#21810) Washington, D.C.

Weiner, E. (1997) *Urban transportation planning in the United States: an historical overview* (DOT-T-92-24), Washington, D.C.: U.S. Department of Transportation.

Notes

1 The debate involves consideration of the various modes of transportation from low-cost infrastructure for individuals (trails and bike paths) to developing expensive rights of way and bridge infrastructure for the automobile, truck, bus and mass-transit/rapid-transit (metro-bus, metro-rail, trolley/light-rail) including the introduction of smart highways.

2 The Washington metropolitan region is being reclassified as a severe non-attainment area for Federal air quality standards. Unless the region reverses the trend, Federal funding can be withheld from transportation projects.

3 This metropolitan region represents a significant segment of the Chesapeake Bay watershed. Residual chemical pollutants from urban sprawl and outlying agriculture usage is seriously threatening the delicate eco-system and aquatic economy of the bay.

4 See the Washington Council of Governments web site www.mwcog.org for information on the roles and responsibilities of local governments in the public decision-making process.

5 In discussions with the George Mason University, the Institute for Conflict Analysis and Resolution (ICAR) transportation study team heard from the longtime mayor of the City of Fairfax, who indicated that 'the predecessor to WashCog actually began regional transportation work as early as 1937'.

6 See Chapter One, Figure 1.1 *The Civil Society: An Interactive Model of Participatory Governance*. This interactive model of participatory governance represents the three sectors involved in the public decision-making process: *government*, *associational* and *business*.

7 Interstate highway I-95 is the major north-south, interstate highway on the East coast (the Florida Keys to Maine) and therefore serves the economic interests of more than just the Washington metropolitan region. The Wilson bridge over the Potomac River is owned by the Federal government, an unusual situation. Other transportation infrastructure is owned by either the State or local governments. The Potomac River is under the jurisdiction of Maryland, but ownership is generally joint to the center of the river channel. The construction and maintenance of the bridge, however, is a joint project of the U. S. Department of Transportation, the State of Maryland and the Commonwealth of Virginia cooperating with the District of Columbia.

8 *Transportation*: personal automobile – roads and bridges versus mass transit – light rail and rapid bus, etc. *Land-use*: a public resource for community and high-density uses versus land as a resource for personal development and economic gain, etc. *Environment*: degradation of the metropolitan air quality and protection of the Chesapeake Bay catchment area, etc.

9 George Mason University is a Commonwealth of Virginia state university at Fairfax, Virginia, USA.

10 The author of this chapter was one of a four member ICAR transportation study team at George Mason University. He also conducted the three-round ICAR Delphi study. The ICAR faculty advisor for both the study team and the Delphi was Frank Blechman. The other members of the study team were Tim Sueltenfuss, Joy Kramper and Keisha Hargo.

11 Starting with a small core group of known public leaders, the number was expanded by the use of a social-metric reputational technique, whereby the initial respondents nominated a number of additional leaders in their own and from the two other sectors.

12 The Delphi, a process for conducting a virtual conversation was developed by the Rand Corporation think-tank during World War II. As adapted for the transportation decision-making project it allows busy officials to discuss, share opinions and develop policy options when it is not possible for them to attend one or more lengthy meetings. The technique comprises two or more questionnaires with each succeeding instrument being developed from the responses of the previous questionnaire. Care is taken so that each succeeding set of questions represents a dialogue which builds on the participants preceding responses. This technique allows busy officials an extended dialogue with other busy officials without the time required to attend meetings.

13 This method of lightly summarizing the responses from an open-end format made for a rather lengthy second Delphi instrument. Since it was a dialogue, it was thought best to share the richness and texture of the previous responses.

14 It was easier to consolidate the responses from the second Delphi than the responses from the first round. A weighted summary of the mainly closed-end responses was used to assign a value that was based on the position of the response along a scale which measured the respondent's affirmation and challenge to each component of the Delphi generated dialogue.

15 The number of groups represented (89) was slightly smaller that the number of respondents (over 90) because one organization was represented by a state and by a local chapter.

16 During the Delphi an attempt was made to introduce 'collaboration' as a different concept from 'coordination' and 'cooperation'. The intent was to introduce a more shared process for resolving differences. Collaboration represents a desire to work together by sharing resources to achieve a common goal. Coordination represents a desire to partner, but not share resources, to achieve a common goal. Cooperation represents a sharing of ideas but

share resources, to achieve a common goal. Cooperation represents a sharing of ideas but not work or resources (Hughes *et al*, 1998; Mattessich and Monsey, 1992).

17 The Commonwealth of Virginia elected top State offices in 2001. Maryland and the District of Columbia in 2002. New governors were elected in Virginia and Maryland and the incumbent returned in the District. In the Virginia elections several local elected officials who supported moderate growth policies were defeated by candidates favoring even slower growth policies.

18 During the legislative debate on granting Northern Virginia the referendum authority for their citizens to approve the creation a special taxing district to fund transportation projects, education advocates lobbied to include funding to meet educational needs, but that failed to gain acceptance.

19 In an essay, Robert Fishman (Katz, 2000) said, ‘A [new governmental] *region*, someone has wryly observed, is an area safely larger than the last one to whose problems we found no solution’ (p2).

Chapter 8

Conflict Management and Collaborative Problem-Solving in a Protected Area in Ghana

David Deshler and Kirby Edmonds

Introduction

This chapter describes an adaptive collaborative management case that evolved from a serious conflict between several villages and guards of a protected area in the Greater Afram Plains (GAP) in January, 1998. The collaborative management process successfully averted armed conflict over the use of the protected area. This sort of conflict is pervasive in protected area management, certainly in Ghana and throughout the world. In Ghana, as a result of desertification, the movement south of displaced subsistence farmers is placing pressure on protected areas. The farmers seek places to settle and practice slash and burn farming techniques that both worsen the desertification and undermine national efforts to preserve forests and wildlife. Furthermore, wildlife poaching is a serious concern in Ghana, as elsewhere in sub-Saharan Africa, and the Ghana government is engaged in armed struggle with organized poaching operations in several places to enforce restrictions on the use of protected areas. This is the environmental context that settlers in and around protected areas find themselves and, thus, the potential for conflict to escalate to violence is high. In addition, because the authority of the central government is evolving vis-à-vis the traditional authority of chiefs, traditional conflict approaches, though important, have limited effectiveness.

This chapter provides a narrative example of the efficacy of well-managed workshops for fostering the type of social learning that can lead to conflict mitigation in the context of protected areas. Two key issues common to these conflicts are addressed: first, access rights for local residents in and around designated parks and preserves; and second, the necessity of and benefits from assuming a pluralist approach that accommodates the interests of multiple stakeholders. The authors were involved as design consultants and observers of the workshop process and as technical experts in the area of natural resource preservation and conflict management.

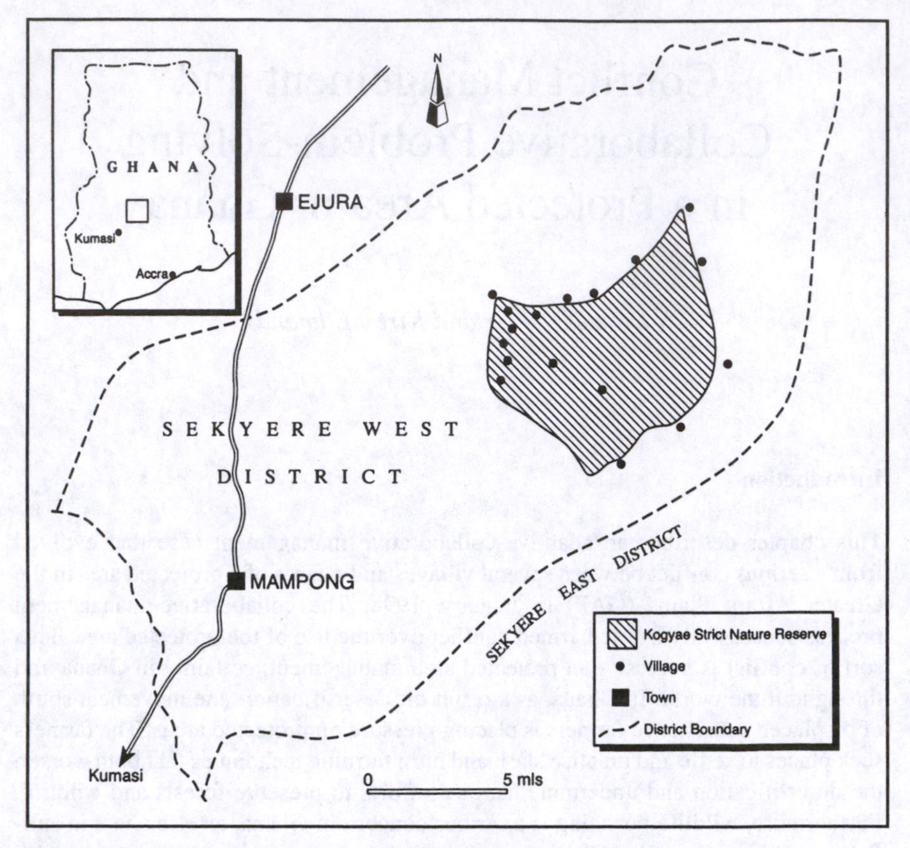

Figure 8.1 The Locational Context of Kogyae Strict Nature Reserve

The situation

The conflict between the guards of the Kogyae Strict Nature Reserve and the people living on the periphery of the reserve in Sekyere West District has been longstanding, has slowed development in the area, and has been a major concern for the District Assembly, World Vision and residents in the reserve. There are eight village communities currently sited within the Kogyae Reserve (Figure 8.1). These are Domi, Atakpami, Sankasase, Dagomba Nyamebekyere, Asase Konkonba, Aberawanko, Sasebonso and Yaya Akura. Village communities sited along the fringes of the reserve are Domi Balana, Taylor Akura, Congo, Medira, Oku Junction, Yaw Moammre, Berem, Kyenase and Kyekyebon. Some of the people in these communities settled in the area many years before the demarcation of the game reserve, while other settlements developed after the demarcation. For example, Sasebonso is one of the villages, which was already situated in the area before its demarcation as a game reserve.

To compensate the community, the Forestry Department has given a land concession to the farmers, but there had been agitation for more land. On the other hand, people of Asase Konkonba community have been asked to leave since they settled in the area after the demarcation of the Game reserve. The demarcation of the protected area has created tension which led to the crisis that will be described below. The Game reserve officials began putting pressure on the people to move out of the area and resettle elsewhere after it was discovered that the chief of Dagomba Nyamebekyere had, alone, farmed about 10 square miles within the reserve. Also, it had been observed that the slash and burn farming practices in the reserve had been contributing to bush fires and environmental degradation. Consequently, the people in the reserve were given official notification to leave the area.

An exacerbating factor was that the paramount chief, who originally gave permission for the villages to be settled, had recently sold some land to the Department of Wildlife. The Department of Wildlife, having bought the land, thereupon extended the borders of the Kogyae Strict Game Reserve and notified the farmers that they were to vacate their villages. The villagers were unaware of this sale and did not realize, nor believe, that they were living there illegally. World Vision has been a stakeholder to the conflict because, although advised not to do so by Wildlife Department officials, it had, with the approval of the District Assembly, the Department of Health and the Department of Community Development, assisted villages inside the Kogyae Game Reserve with the provision of boreholes for potable water. The location of some of these boreholes gave the impression that World Vision was encouraging permanent settlement in the reserved area. In addition World Vision staff had been assisting farmers with cashew cultivation experiments to provide an alternative to slash and burn farming. The boreholes and the cashew experiments signaled that the farmers would be permitted to stay and defused some of the tensions in the short term. But the boreholes and cashew experiments also raised the settlers' stake in permanent occupation of the land which clearly exacerbated the conflict with the Department of Wildlife.

It was clear from farmers' statements at district workshops designed to raise awareness of environmental issues that there was confusion about the status of settlers and that there was conflict between the settlers and the paramount chiefs in relation to security and tenure. Some of the key contextual issues included: boreholes and their contribution to encouraging people to stay, which had been raised by the Wildlife officials; the extension of the Strict Nature Reserve boundary which designated many of the villages inside the restricted area; the degree of exclusion of those in the actual area, particularly the District Assembly, in the way the extension was drawn; and inconsistent enforcement. It was also clear that failure to reduce these tensions and resolve these conflicts would have serious implications for the future well-being of the people, as well as for the environment that the guards were seeking to conserve.

Conflict management workshop proposal

Awareness of these issues led to a proposal to conduct a conflict management workshop to prepare people in the area to deal with the issues in a way that would prevent the conflict and tensions from escalating to violent confrontation between the farmers and wildlife guards. Initially the intention had been to work with the academic community in order to build a foundation for understanding the nature of the conflict, to blend traditional Ghanaian approaches to conflict management with other approaches, and to prepare people to think about how to handle this and other conflicts that might arise. However, this workshop was delayed as information emerged that the conflict in the Kogyae Reserve was quickly moving beyond the preventive stage.

A key event was the seizure by the wildlife guards of the tractor of the best farmer in the area. An award he received for innovative farming practices and productivity brought to their attention the fact that his farm was located in the reserved area. They confiscated his tractor and, in his outrage, he made some injudicious and threatening remarks to the guards which ultimately led to his arrest. Furthermore, when the District chief executive visited a village to help resolve this issue none of the farmers came to the meeting that he had called. This made it clear that the situation was deteriorating and becoming more serious. Wildlife staff were viewing the matter of power and rights so differently from the community that traditional methods of arbitration seemed inappropriate for handling this conflict.

During November, 1997, Dr. Kwesi Opoku-Debrah, a Cornell Senior Extension Associate, was informed by World Vision that the conflict situation at the Kogyae Strict Game Reserve in the Sekyere West part of the Greater Afram Plains (GAP) was intensifying. Farmers had responded to the increased pressure from the Game Reserve officials by arming themselves and threatening to shoot game reserve guards if they trespassed on their farms or villages. Dr. Opoku-Debrah responded to this situation by negotiating a workshop among all the stakeholders which was held at the Ministry of Agriculture College for Extension Staff at Ejura. Additional housing was provided at the Ejura World Vision Resource Center which is nearby. Both facilities are approximately 20km from the reserve. Approximately 70 participants were invited and attended the workshop, including District officials from Sekyere East and Sekyere West.

Identifying and engaging key stakeholders

A meeting was held between the Natural Resource Management and Sustainable Agriculture Partnership (NARMSAP) and the Zonal Committee which includes chiefs, and elected community representatives (assemblymen and assemblywomen) to discuss their willingness to hold a conflict management workshop. The interest they expressed, both in having the workshop and in being participants at the workshop, underlined their level of concern and commitment to addressing the conflict directly. At this meeting it became clear that World Vision was being urged to take on the traditional role of arbitrator. World Vision carefully refused this role. It was agreed instead that World Vision, rather than acting as a judge or arbitrator, would act as a convener and facilitator at this workshop which would include all of the stakeholders so that each could tell the other their stories and work on solving the problem together.

Hence, a 5-day conflict management workshop was scheduled for mid-January 1998, in an attempt to find solutions to the issues. The operational objectives included: creating a climate for listening and learning from each other regarding ethnic conflict and conflict in protected areas; facilitating stakeholder participation in the identification and assessment of the practicability of traditional (indigenous and foreign) conflict management approaches; planning effective and sustainable implementation strategies to be adopted by legitimate stakeholders; documenting the nature of the conflict and recommendations agreed by stakeholders to resolve the conflict.

About seventy participants, comprising officials from the Sekyere East and Sekyere West District Assemblies, Cornell International Institute for Food, Agriculture and Development (CIIFAD), representatives from the Friedrich Ebert Foundation and interested donors, and World Vision staff as well as other stakeholders in the area, were expected to participate in the workshop. Calling the event a "workshop" rather than "mediation," or a "conflict resolution meeting," made it more neutral to all participants. The people involved, including the farmers, were familiar with NARMSAP workshops and thus were more relaxed about participating. World Vision field staff went to each of the villages to personally deliver invitations to the community representatives and to urge them to come. World Vision also took responsibility for arranging and providing their transportation. CIIFAD provided funding for travel by village participants and accommodation for all participants during the workshop.

Media presence was determined to be advantageous because the media had expressed interest in this issue. They were told that they would be included in the opening and closing ceremonies. It was important to include the press in a formal way because legally, the District Assembly must give the press access to its activities. At least in this case, the media were very constructive, lent dignity to the proceedings, and their presence helped encourage constructive efforts. Rather than posturing or grandstanding, participants tried to be reasonable, knowing that their words might be reported on the evening television news.

World Vision took responsibility for securing the participation of the Wildlife Department staff. There were difficulties in getting direct communication with the Director but there was initial indication of interest and intention to attend the workshop. An invitation was also extended to all of the guards. These initial difficulties were probably due to the high stakes and the fear being experienced by the guards, but this did not become clear to the workshop organizers until the day before the workshop was to be convened.

By Monday evening, members of the District Assembly, village chiefs and farmers had arrived, along with World Vision staff. However, only one staff member from the Department of Wildlife had arrived and the workshop was scheduled to begin formally on Tuesday morning. After looking at the agenda the lone wildlife officer was relieved, but said that the others were not present because they feared for their lives; the farmers had armed themselves and had threatened them. He said that, while he was not authorized to order the others to come, given the agenda, and World Vision's assurance of their safety, he would ensure that the guards would arrive the next day. But when the guards had still not arrived by early Tuesday morning, efforts were then made by Kwesi Opoku-Debrah to contact the Senior Wildlife Officer of the Department of Wildlife in order to respond to these concerns. Dr. Opoku-Debrah negotiated with the Senior Wildlife Officer via phone. The guards were reluctant to come without their weapons fearing entrapment by the farmers, but eventually they were persuaded to come unarmed. As a result of these protracted negotiations, the guards did not arrive until 9:30 am or so on the first morning. They came in looking very anxious and the tension in the room was very high since their late arrival had served to raise everyone's anxiety. As for the farmers, a World Vision vehicle and driver had been provided so that they would not feel threatened by the guards arriving in their official vehicles.

The workshop was ultimately attended by district officials from the Sekyere East and Sekyere West Districts, representatives of the District Assemblies, World Vision staff, local chiefs, representatives of paramount chiefs, staff of the Department of Wildlife, The Fire Service Department, Operation Halt program of the Department of Forestry, CIIFAD staff (Kwesi Opoku-Debrah, David Deshler and Kirby Edmonds), and representatives from national television and the national press. An important note about the participants is that many of the farmers and community representatives were Muslim, observing Ramadan that week and fasting during the day. This required that World Vision staff had to provide meals for them at 3:00 am.

Selection of venue

Initially the location of the workshop was to be Oku Junction. That site was rejected because the road was not good and it was too close to the area in dispute. Indeed, a field visit to the area in dispute was thought initially to be a design advantage but, because there was too much suspicion, this did not happen at the outset. After the opening ceremony the media expressed interest in going to the area in dispute. There was some conflict with local community representatives who expressed their suspicions about why the media wanted to go to their villages. This was resolved and the media did take footage of the workshop, the villages, and the wildlife reserve.

Ultimately, Ejura was chosen as the workshop site in order to achieve greater neutrality by getting further away from the area in dispute. The community center was then identified as a possible venue for the workshop because housing was being provided at the Ejura World Vision Resource Center as well as at the Ministry of Agriculture College. Both facilities are approximately 20 km from the reserve. The agricultural college was finally chosen as the agreed venue for the workshop when the facilitators realized that the community center was too small to accommodate a group of over 75 people.

Facilitator preparation

The workshop design team comprising three CIIFAD representatives (Kwesi Opoku-Debrah, David Deshler and Kirby Edmonds), two additional World Vision staff including the Asare Boateng, and the World Vision manager of the Ghana Rural Water Project (Emmanuel Opong), met on Sunday afternoon before the workshop to make their arrangements. It was at this meeting that Asare Boateng and Kwesi Opoku-Debrah were selected as the primary facilitators, partly because they already had good relationships with the farmers, but also because both had excellent facilitation skills and were very good translators of several languages including Twi, English, Akan and Ewe. Because the workshop would be conducted in Twi, it was determined that the best role for David Deshler and Kirby Edmonds (both English speakers) would be as process observers and consultants. Each would receive whispered translations as the workshop proceeded. An agenda was drawn up to include the following activities: introductions by all participants; workshop norms; opening ceremony; expectations for the workshop; historical timeline; definitions of key words in the conflict; current use of the natural resources; possible scenarios for the future; consensus on criteria for acceptable options; reports on CIIFAD, on conflict approaches, and by the Departments of Wildlife and Forestry; resolutions and agreements including next steps; and closing ceremony.

Workshop activities: Tuesday

Introductions: On Tuesday morning when the workshop was scheduled to begin, even though the Department of Wildlife staff still had not arrived, the facilitators began by having those present begin introducing themselves. This allowed each of the participants present an opportunity to speak, identify his/her affiliations and hear from the others. It was unfortunate that the Wildlife Department staff were not present for this activity. At approximately 9:30 am, the guards arrived, marched in, and took their places on the west side of the assembly area. They introduced themselves and participated in setting norms for the workshop. The most senior wildlife officer in attendance during the workshop arrived later in the day.

Workshop norms: This activity was important in establishing some agreements for behavior throughout the workshop and to give participants a chance to hear from each other (particularly since the guards had arrived late). The group brainstormed a list and after some discussion agreed on the following norms which were posted on a flipchart for all to see:

- Respect your friends' views.
- Raise your hand for any question.
- No smoking/drinking.
- Be patient for discussions.
- We should be timely.
- No sleeping.
- No interruptions.

Expectations: For this activity, people met in stakeholder groups (including chiefs, Wildlife Department staff, NGO staff, District Assembly representatives, community residents, and media representatives) where each person would be fairly comfortable, These groups were asked to identify their expectations for the workshop. Each group expressed its hopes and fears, and was able to communicate these without comment or discussion when their ideas were recorded on newsprint sheets and hung on the walls as group reports. Happily, most of the groups expressed an expectation (hope) that some peaceful resolution could be found. This activity was interrupted by the official opening ceremony. The scheduling constraints of the dignitaries for the front table had prevented their being present at the beginning of the day.

Opening ceremony: The opening ceremony was video taped and recorded by the media including Ghana Broadcasting Corporation, Chronicle, Graphic, and New Times corporations. It included prayers and speeches from District government officers, the national director of World Vision and others. The opening ceremony was an important ritual that formally recognized shared leadership for this effort and showed to all that the workshop was being observed and regarded as important by many people in the country who would not be present to participate. Given the importance of religion in peoples' daily lives, the prayers during the opening ceremony, asking for guidance, wisdom and a peaceful solution, helped set a positive climate for moving forward.

Expectations (continued): The reports were written in English and the facilitators translated them verbally into Twi. The activity generated an array of issues and interests from participants. It let facilitators and other participants hear everyone's expectations and gave the facilitators an opportunity to identify whether there were any expectations that they were not prepared to address directly. It also gave participants an opportunity to identify shared and conflicting expectations. Having the media participate in this activity underscored the public visibility of the collective effort. The news media and television reporters, after the dispute mentioned above, went with community representatives and a World Vision

Table 8.1 Expectations by Stakeholder Groups

NGOs	Media	Farmers from the Communities
To achieve a level of understanding and appreciation of the conflict issues among all stakeholders as regards the conflict; WVI has made major investments of money and skills. To acquire knowledge about conflict prevention. To know what WVI has done so far. To harmonize the views and interests of all stakeholders. To focus more on the social aspects of managing the nature reserve To know what lessons can be learned from this conflict and apply them elsewhere. To ensure that relationships are established among various groups to bring absolute peace in the area. To learn about how sustainable development can take place. Establishment of international relationships with regard to conflict management. Establishment of the process of conflict management	Humane approach should be adopted to solve the problem. Some of the existing laws and regulations should be reviewed. WVI should liaise with government institutions dealing with game, wildlife and lands. Commission to know the no-go areas and siting of bore holes. Farmers should also review existing farming practices. NARMSAP should go beyond the role of advocacy by educating farmers on modern farming strategies. Traditional rulers should also be made to see the dangers in allowing the settlers to farm in certain areas. Alternative crops like hybrid rice and cashews should be introduced to the farmers as against yam cultivation.	They should make a new demarcation for us. The wildlife staff should not harass those inside the reserve. There should be peace between the communities and the Game and Wildlife staff.
Chiefs	**Department of Wildlife**	**Wildlife** (at the urging of one of the CIIFAD observers, one of the wildlife staff wrote a set of expectations on behalf of wildlife)
They should make a new demarcation for us The wildlife staff should not harass those inside the reserve. There should be peace between the communities and the Game and Wildlife staff.	Better community understanding and cooperation with the Wildlife Department in the preservation and conservation of resources of the Kogyae Strict Nature Reserve.	We need peace in our environment to stay and breed for your future survival. We have the right to life as you do, please stop harassing and killing us. Stop destroying me for I give you shelter and protect your water sources for your survival and give you rains. Remember the role I play in making your land fertile. Stop the burning which destroys my habitat.

communications officer, to the Kogyae Strict Nature Reserve to take photographs of the reserve. The news story was featured in newspapers and on national television that evening. The expectations of key stakeholder groups are summarized below in Table 8.1. One of the most significant things that emerged from this activity is that each party took the opportunity to acknowledge in some way the importance of a resolution that included both the presence and the needs of the other parties. This, along with the conciliatory tone established by the opening ceremony, allowed everyone to relax somewhat and reassured them that there was probably room for movement from positions that had previously seemed intractable.

Lunch: Participants took food during the lunch-break to their stakeholder seating sections of the assembly area. There was little interaction among stakeholders during lunch on this first day.

Timeline: After lunch, the workshop participants spent the entire first afternoon on a timeline activity for the region now known as the Kogyae Strict Nature Reserve. This involved each of the participants in a personal way in the stream of events leading to the present situation. Everyone took part in describing the events that occurred in the Kogyae area beginning from around 1920 to the present. The activity made it possible for each stakeholder's involvement with the area to be acknowledged publicly and gave every participant a better understanding of how the situation was being viewed by others and a greater appreciation for the varied perspectives on the history of the region. The timeline put everyone on an equal footing, and even gave those who had the longest acquaintance some "seniority", regardless of their status. It also created a shared group history out of the individual histories. It enabled people to learn more about one another and created openings for informal dialogue between participants. Participants were given 3 x 5 Post-Its to identify themselves by name and to record events such as when he or she first became associated with or came to know about the Kogyae reserve, his or her connection to it, and events from their lives associated with their activity in the Kogyae area. Again, participants sat in stakeholder groups which was important, not only for the sake of familiarity and the validation of experience, but also to make it safe for participants to remind each other of events that had occurred in the region. A long strip of flipchart paper was taped to the wall, running the full length of the workshop room. The Post-Its were written in English and in other local languages. Non-English speakers and non-writers were assisted by others in their group. Each Post-It was read to participants both in English and in Twi. The Post-Its were typed by the next morning so participants could have a hard copy of the timeline.

Dinner: Participants took food during the dinner period to their stakeholder seating sections of the assembly area. There was little interaction among stakeholders during dinner on this first day. After dinner some members of the workshop watched video replays of the day's proceedings. This turned out to be significant because having planned no other evening activities, watching the videos provided an opportunity for guards and farmers to sit with each other informally.

Workshop activities: Wednesday

After morning devotions and breakfast on the second day, facilitators asked participants to review the significant activities from the previous day. There was general agreement that the timeline was important in establishing that everyone present had a legitimate stake in the area and that they all had something to learn from each other and something to teach.

Definitions: The remainder of the morning was spent in defining the meanings of words and concepts associated with the conflict including: forest, reserve, game, wildlife, settlement, conflict, ownership and extension (ie. new boundary). While designed as a 'neutral' exercise, it turned out to be exceptionally enlightening as participants came to understand and acknowledge the different meanings that stakeholders held for the same terms. Attempts by stakeholder groups to translate terms such as 'forest' or 'reserve' into local dialects indicated how misunderstandings can and had occurred through different interpretations. The term 'reserve' for example, was translated into Twi with the word for 'sacred grove' because nobody is supposed to go to the reserve and no one is supposed to go to a sacred grove. But to farmers, this made little sense for the strict nature reserve, about which there was nothing considered sacred.

This exercise helped to draw groups together. The activity also served to equalize power by crediting the knowledge, language and wisdom of local people. It helped to establish a common language for the group and to illuminate underlying assumptions and worldviews embedded in the words, particularly across languages. During this activity stakeholder groups built on their shared views, experiences and definitions. The reports from the stakeholder groups allowed facilitators to begin articulating an emerging understanding in the group about the nature of the misunderstandings and the sources of some of the conflict.

Resource assessment: The afternoon was spent identifying natural resources in the area using stakeholder groups. The exercise focused on uses of natural resources, their present state, conservation practices, and current destructive activities. Trends and prognoses were also discussed. The deliberations provoked remarkable candor, even becoming somewhat confessional. Forest guards, for example, admitted that they did, perhaps, carry on some poaching once in a while. The result was to make clear to all that present resource uses were neither ideal nor sustainable. Each stakeholder group posted its assessment of resources for all to see. Flipcharts were scribed in English and translated orally to Twi. It was important to do this activity in stakeholder groups, because this provided safety that allowed for honesty. Thereafter, the reports to the large group led to a shared recognition of the threat to the resources posed by current practices and laid the foundation for a common and shared understanding of the nature of the problems.

Dinner: There was a little more mingling among the participants at this meal. After dinner, some participants reviewed the workshop events by watching that day's videos.

Workshop activities: Thursday

Presentations: On Thursday morning, after a review of activities, brief presentations were made by resource persons. These contributed ideas that could help to explain, rationalize or justify new resource use practices and new methods of cooperation. A member of the design team from CIIFAD added context to the meeting by lessons being learned in various places in the world concerning mechanisms for sustaining protected areas. In addition to a legal approach which sets aside land and generates laws, and the scientific approach which helps us to understand the limits of ecosystems, he asserted that recognition should be given to ecological justice, environmental economics, and spiritual and cultural foundations that sustain an environmental ethic. He emphasized that participatory management was being tried in many other places and briefly described how it could work.

Another design team member from Cornell University reminded the group that effective conflict resolution requires that all stakeholders be willing to learn about common interests and ways to conserve without destruction. He also pointed out that learning could be difficult for adults because deciding to learn implied a recognition that we could be ignorant or in error. He emphasized the importance of remembering the stakeholders who are not present in this workshop and taking home a consistent account of this workshop to those who had not participated and would still be feeling the mistrust that participants in the workshop had managed to overcome.

A participant from Operation Halt in the Forestry Department explained the need to limit farming and prevent further deforestation. He pointed out that historically in Ghana, farmers who have lived in reserve areas have been allowed to stay, but new settlers have not been allowed. The failure to enforce this gave rise to Operation Halt whose activities focus on eliminating poaching in reserved areas and eliminating illegal settlement. A senior Wildlife Officer, recounted efforts to protect the natural area, beginning in 1970, indicating that communities did not cooperate and that settlements have expanded. He stressed the importance of being tolerant and the importance of establishing a community management committee approach.

Lunch: At this meal participants were much more random in their seating, with guards sitting and chatting with farmers.

Possible scenarios for the future: After lunch, participants were randomly assigned (by counting off) to eight mixed groups to make best guesses about what would happen in various scenarios regarding the current conflict. The scenarios included:

- The old demarcation to be maintained.
- The old demarcation and present extensions to be maintained.
- The wildlife workers to leave the reserve.
- The residents to leave.
- Both residents and guards to stay.
- Resources become depleted.
- Resources are maintained and conserved.
- Both community members and guards socially manage the reserve.

This activity brought participants with diverse interests together to assess the problems and to anticipate outcomes of the possible ways to handle the conflict. Each group reported on newsprint to the total assembly and consensus emerged from this activity on a number of principles and goals as well as acceptable and unacceptable futures and the limits of acceptable options. It was important to use heterogeneous groups in this activity because by this time in the process people were ready to work together. When the facilitators reported the output from the groups, they were essentially reporting on behalf of the collective interests rather than particular stakeholder groups. This resulted in a shared framing of the issues and a common view of the necessary criteria for a mutually acceptable solution.

Resolutions: The scenarios activity was followed by the formation of resolutions which came from the floor. By this time there was a common feeling that 'something must be done', not just to avert violence but to find policies and practices that would support the different legitimate interests at stake. The proposed resolutions, sometimes after a little rewording from the floor, were agreed unanimously. They included:

- The stakeholders present agree that there is a need to readjust the existing boundaries of the Kogyae Strict Nature Reserve to meet the needs of the communities and wildlife conservation.
- There is disagreement over the extent of the boundary adjustment.
- The communities recognize the important roles of the Wildlife Department staff in the area and endorse their continued presence.
- The presence of both Wildlife Department staff and the community members is needed for the effective management and conservation of the resources in the reserve.
- Wildlife Department staff and the communities should work together to enforce existing wildlife regulations and make by-laws whenever necessary.

- A Management Committee should be formed to draw up a Plan of Action that would include immediate implementation of the resolutions, boundary demarcation planning, and relaying these resolutions and action plan to the national government.

By this point in the workshop, the entire group was working in concert to arrive at mutually acceptable solutions that would allow the group to assert that they had achieved a workable direction for the future. Agreement was reached on the composition of a Management Committee comprising 13 persons: 3 village chiefs, 4 community representatives, 3 Department of Wildlife staff members, 2 District Assembly officers, and 1 NGO representative. Each stakeholder group met and named its respective committee members.

Closing ceremony: The closing ceremony included prayers, acknowledgments, and speeches regarding the importance of conflict resolution and patience in the days to come. Workshop participants were in a much relieved mood for dinner that evening and for breakfast the next morning before leaving. They mixed with each other and were eager to have a group photograph taken of all participants.

Post workshop activities

Shortly after the workshop, the Management Committee convened and established a budget and a bank account for its work. Costs were shared among the Wildlife Department, Sekyere West District Assembly, and World Vision. The communities promised volunteer labor. A reconnaissance trip to the affected communities was undertaken and a stakeholders meeting was held on February 26th 1998 at the District Assembly Hall at Mampong. The Management Committee discussed with the communities their expected roles. A draft memorandum of understanding (MOU) for the management of the proposed Special Use Zone (SUZ) was drawn up and submitted to the District Assemblies, the communities, and the Wildlife Department. Pending the final approval of the MOU document it was agreed that:

- The Special Use Zone would be recognized as part of the Kogyae Strict Nature Reserve.
- No new farms would be cultivated beyond the newly adjusted line.
- All farming beyond the new line would cease after December 1998.
- The communities would cooperate with the Department in the management of the reserve.

With the help of a surveyor, the village boundaries were redrawn with all the committee members participating including the wildlife officers. This did not occur without difficulty. In spite of the communities agreeing to the proposed adjustments during the various meetings, they refused to cooperate when the actual work began in the field. Some villages insisted on the line being shifted to the original boundaries of the Kujani Bush Reserve and this brought the work to a halt for two weeks. After a series of meetings facilitated by World Vision staff, compromises were made which enabled the work to resume.

In addition, the promised labor from the communities was not easily and readily available and the Department of Wildlife had to provide the needed labor to cut the lines. Though the communities had agreed to provide meals and lodging for the surveyor and his assistants to reduce costs, they did not do this. Furthermore, the surveyor and his team were frequently intimidated in attempts to force them to agree to the wishes of local people; in one community some of the erected pillars were removed and destroyed. However, members of this community immediately identified the culprit and police action was taken against him. There were a few isolated cases where new farms, charcoal burning, hunting, and tree felling were challenged by members of the communities and reported to the police. This phase took sixty-six days after which a celebration was held on May 20th 1998. It was designated as the 'Untying of the Knot Celebration'. Participants from all the villages gathered with the wildlife officers to hear speeches, dance and to eat together to commemorate the agreements and to affirm their resolve to follow the rules to which they all had agreed.

The collective MOU for the SUZ was also revisited. The agreement included the following:

- Maximum of four acres of farmland to be cultivated by each farmer within the SUZ.
- Total ban on felling of timber tree species during farming and anywhere in the SUZ and reserve.
- Non-encroachment in the reserve.
- Total ban and cessation of charcoal burning in the SUZ. A ban on palm wine tapping. Any palm-wine tapping would be carried out only in consultation and agreement with the Wildlife Department staff stationed at Dome.
- Ban on bush burning and group hunting.
- Removal of all food crop items in the reserve by the end of February 1999.

The CCMC further explained that the SUZ was still under the control and management of the government i.e. the Wildlife Department. The SUZ had not been devolved yet to the communities but Parliament, by Legislature Instrument, could facilitate this. The process, it was explained, was a difficult one and could only be achieved if community members remained tolerant and law abiding. The communities were assured that the stakeholders in the process (WVI Ghana, District Assembly and the Wildlife Department) would jointly pursue the issue to its logical conclusion by passionately appealing to Government to devolve the SUZ for entire community ownership and use.

Meanwhile community members were urged to expose and arrest a few farmers amongst them who were continuing to disregard the agreement. Unit committees were to help arrest such offenders for prosecution by the courts and were to help monitor the four acres farmlands. Community members were also urged not to quarrel or fight with the Wildlife Department staff because they were engaging in their assigned profession.

Minutes of a Management Committee meeting during the Fall of 1998 indicate that in all villages inside the wildlife reserve, the interactions among the major stakeholders had been very helpful and successful. Attendance and participation at the Unit Committees were reported to be high, and useful suggestions and comments were made by all the communities regarding the effective management of the reserve and the SUZ. At the Management Committee Meeting the following villages gave reports as follows:

Yaya Akura: 'Thanks for all your efforts. We are glad we have now been permitted to stay in our village. We pledge to be law abiding so that we do not disturb the present peace. Our problem is that we do not have potable water and the only available stream for use is in the Reserve. WVIG should kindly help us with boreholes. Meanwhile, we promise that even if we should enter the forest, it would be only to fetch water and nothing else. We shall monitor the smooth implementation of the MOU'.

Aberawanko: 'We live very near to the Reserve. We have a few farms in the reserve. We will remove all our food crops from the reserve by the end of February 1999. We are appealing for tractor support from the Wildlife Department'. They also appealed to WVI Ghana Rural Water Project (GRWP) to put a hand-pump on a borehole that was drilled some time ago. They explained that although a new demarcation had been made, their normal tractor vehicle road still passes through the reserve. The wildlife officer explained that very soon the road would be diverted but in the interim, they should behave in accordance with the agreement whenever they use the road in the Strict Nature Reserve. A WVI official also assured them that he would carry the concern expressed on the borehole to WVI GRWP Water Resources Department for action.

Kyenase: community members suggested the need to have training in land use management as well as ensuring effective boundary demarcation between communities and community farmlands. This, they explained, would forestall any conflicts with adjoining community members.

Sasebonso: the local leader was full of praise for the committee's work and regretted the few incidents/breaches that had happened in his area. He was of the hope that these violations would not be repeated and pledged his full support to monitor the peace of the area. He also pleaded for a borehole from WVI GRWP.

Asase Konkonba: The local leader also expressed regret about the encroachment incident by some members of his community. He promised to bring out the culprits, but pleaded that they should be dealt with leniently. He would ensure that these incidents would not be repeated. They would abide by all the MOU decisions.

Dagomba Nyamebekyere: The local chief expressed support for the CCMC and expressed the hope that his people would respect to the letter all the decisions in the MOU. He commended the CCMC for the way it had managed a misunderstanding between one of his community members and another community member at Dome and expressed hope that the incident would not happen again. He asked for a school to be provided for the children in his community. The Assembly member and chairman of the CCMC confirmed the Assembly's determination to build a school for them. They were asked to prepare some bricks.

Domi: The Unit committee confirmed that they would visit various farms to ascertain the four acres farmlands purported to be cultivated as agreed in the MOU. They would also assist the Wildlife staff in monitoring all activities in and around the reserve.

The minutes ended with the following statement: 'The educational program has been a very useful one because it has enabled committee members to respond directly to community responses and concerns under the new dispensation. It is gratifying, however, that all community members living along the fringes of the KSNR are determined to live there peacefully without encroaching or committing acts likely to affect the reserve and the SUZ. There have been a few challenges, but it is now hoped that with time, things will settle in their right places'.

Meanwhile the Wildlife Department explained that a major management plan for the area would soon get funding so that communities would be assisted with some facilities and skills that would enable them to be less dependent on the resources of the reserve and SUZ. The Department assured them that by the end of the year the program would be in place and residents were advised to be law-abiding and help to preserve and manage the reserve for posterity. The next meeting was scheduled for mid-March 1999. Between that fall meeting and the March meeting, farmers made trips to the University of Science and Technology at Kumasi to seek training in sustainable agricultural practices.

Lessons learned

It is not easy to transform near armed conflict among stakeholders into collaborative management of a protected area. In this case, however, it became possible. Traditional approaches to conflict resolution that make use of those in charge to mediate conflict appear to be insufficient. This is because both local and paramount chiefs were themselves major stakeholders in the natural resource and protected area disputes, both among themselves and with government. In addition, the government units at various levels were not in a position to mediate because they too were major stakeholders. This case illustrates the potential role of NGO's in a conflict management situation that led ultimately to collaborative resource management outcomes.

From that experience the first lesson learned was that many disputes must be transformed into spaces for learning. The presence of a tradition of workshops among the farmers, sponsored by NARMSAP created a familiar space for learning. The workshop format assumed a 'learning our way out' stance to problems and recognized that local knowledge, external knowledge, and the available resources of the partners could create solutions without have solutions imposed from the top down. Past workshops had generated substantial social capital that allowed a team of university and NGO facilitators to bring the stakeholders to the table. NGOs that have earned credibility, along with university support, may be able to provide conflict management leadership for some protected-area disputes if they take a democratic participatory approach. In addition, resource persons from outside Ghana may have been helpful by increasing the motivation (particularly among government officials) to find fair settlements.

The process activities used in the Ejura Workshop appeared to work for the stakeholders. Particularly helpful were the activities, beginning in stakeholder groups, that established a common history, definitions of key words, analysis of current resource use, and the testing of future scenarios to identify acceptable long term solutions. The facilitators also were respected and skillful. The use of ritual (opening and closing ceremonies and traditional recognition of conflict resolution such as the "untying of the knot' celebration) can be important assets toward building relationships upon which collaborative management must rest.

The acceptance and fairness of the management structure and full representation of stakeholders on it contributed to the resolution of specific operational issues and the successful challenging of those who were failing to comply with agreements. The sharing of costs for the management structure was also essential.

In summary, we conclude that a legal approach which sets aside land and generates laws for protected areas without collaborative management is clearly insufficient, and may cause armed conflict. It is also insufficient to expect wildlife officers to patrol an area where farmers view them as enemies. Moreover, sustainability is unlikely if the designation of protected areas results in ecological injustice and economic hardship for former occupants. What is required is space for learning and the identification of an environmental ethic that is supported by spiritual and cultural norms on the part of institutions and local stakeholders. Once this is recognized and relationships are negotiated for specific responsibilities, then collaborative management within the context of participatory governance can become a reality.

Chapter 9

Participatory Regional Planning in Northern Ireland

Michael Murray and John Greer

Introduction

At first sight Northern Ireland may seem a strange place in which to find a collaborative, inclusive and discussion-driven approach to strategic spatial planning. Over the past 30 years bureaucratic government has overseen policy formulation and implementation processes on behalf of a society marked by deeply contested histories and cultures, bitter sectarian violence and endless constitutional wrangling. But daunting as those circumstances have been, recent process related innovation within the sphere of physical planning has been able to harness the enthusiasm of locally elected representatives and the energy of an emergent associational sector in the shaping of a new regional framework for economic and social development and environmental stewardship. This chapter is concerned with an analysis of that experience.

In June 1997 the then Secretary of State for Northern Ireland, Dr Marjorie Mowlam, initiated the task of preparing a regional strategy for Northern Ireland. Over the previous 18 months work had been carried out by the Department of the Environment (DoE) on a new strategy for the Belfast City Region but, rather than bring the latter to a conclusion, it was sensibly decided to merge it with the formulation of a more coherent Northern Ireland-wide analysis and plan. In November 1997 a discussion paper titled *Shaping Our Future: Towards a Strategy for the Development of the Region* was published by the DoE to stimulate public debate and to facilitate input to the planning process from multiple interests.

This was followed in December 1998 by the publication of a draft version of the Regional Strategic Framework (RSF) which set out a series of planning guidelines related to:

- strengthening regional cohesion;
- a spatial development strategy for Northern Ireland based a hub, corridor and gateway approach;
- a balanced approach to regional growth with new housing targets to 2010 distributed across the 3 settlement groups of the Belfast Metropolitan Area, Derry/Londonderry and the Regional Towns, and the Rural Community;

- the development of a modern, integrated transport system;
- support for economic development;
- environmental management and the creation of healthy living environments.

A public examination, the first of its kind in Northern Ireland, was convened in the Fall of 1999 to test, through discussion among invited stakeholders, the content of that draft document. The panel appointed to conduct the public examination subsequently reported to the Department for Regional Development (DRD) in January 2000, the report was published the following month (Elliott *et al*, 2000), and further public input on the acceptability or otherwise of the panel's recommendations was invited through to the end of May 2000. In April 2000 the DRD issued its own initial response to the report of the panel to assist interested parties in submitting their final views by the May deadline (Department for Regional Development, 2000). Following lengthy (and confidential) interaction between DRD officials and the Northern Ireland Assembly the regional strategy was eventually adopted in September 2001 (Department for Regional Development, 2001).

The timescale associated with the preparation of the RSF has, thus, been considerable and has required public participation at different stages over a period of 3 years. This style of 'planning through dialogue' is premised on an exhaustive (if not exhausting) search for consensus across a very wide range of stakeholders whose interests, influence and power vary. It is a style of planning which fits well with the contemporary conceptual prominence in public policy of partnership, the associational sector, conflict mediation and democratic renewal. Furthermore, it stands in marked contrast to the expert-invented 'blueprint' regional planning prescription for Northern Ireland in the 1960s and the data-driven comprehensive approach of the subsequent 1975-1995 Northern Ireland Regional Physical Development Strategy - a strategy for technocrats by technocrats. Instead, a prominent feature of the unfolding RSF chronology has been the dialectic of reaching agreement through inclusive discussion on the shape of over-arching policy implementation principles and the geographical representation of these principles out to the time horizon of 2025. This chapter provides a critical assessment of some key elements of the participatory approach associated with policy formulation in this sphere.

The next section deals more fully with the interaction between the evolving shape of policy content and the participation of multiple stakeholders. The chapter continues by reporting the findings from primary research into the opinions of participants at the public examination of the draft Northern Ireland Regional Strategic Framework. The final section explores the contribution of the Northern Ireland regional strategy process to meeting the broader challenges of participatory planning and democratic renewal.

The Northern Ireland Regional Strategic Framework

The Regional Strategy Discussion Paper - November 1997

As mentioned above, the process of preparing a regional strategy for Northern Ireland commenced in mid 1997 and was given significant profile by the publication in the following November of a discussion paper titled *Shaping Our Future: Towards a Strategy for the Development of the Region.* As illustrated in Figure 9.1, the Discussion Paper proposed a daunting (and with the benefit of hindsight, impossible) timetable for completion of the process through to December 1998 and linked with which there were to be considerable inputs by way of public participation. The merging of consultative work associated with the preparation, thus far, of the Belfast City Region strategy into the Northern Ireland Regional Strategic Framework effectively completed Stage 1. That approach was extended during Stage 2 to the whole of the region following the publication of the November 1997 Discussion Paper and, in full, comprised a number of features:

- a total of 116 direct consultations by the Department of the Environment with District Councils, political parties, other parts of Government and regional organizations;
- a total of 207 formal submissions from District Councils, political parties, elected representatives, business organizations and public sector bodies;
- the appointment of a research consortium from The Queen's University of Belfast, The Urban Institute at University of Ulster, Community Technical Aid and Rural Community Network which facilitated consultation with 477 community and voluntary groups;
- the convening of 2 conferences attended by some 600 young people which resulted in the submission of a Northern Ireland Youth Council Report.

Thus on a conceptual spectrum of citizen participation which ranges from information dissemination, to opinion gathering, to active involvement in decision making, through to delegated authority, it is clear that the processes of engagement during Stage 2 were located at its lower end. In contrast, the public examination which formed the cornerstone of public participation in Stage 3 lay more within the domain of active involvement in the policy formulation process, albeit that the setting of the agenda and the subsequent determination of policy preferences were overseen by regionally elected representatives and civil servants.

In terms of content, the regional vision within the Discussion Paper was advanced as 'a better quality of life' to be achieved by 'valuing people, building prosperity, caring for the environment and improving communications'. A central issue was identified as being the provision of dwellings for up to 200,000 households over the next 20 years. Trend analysis pointed to steady population growth during previous decades shared across the more prosperous suburbs of

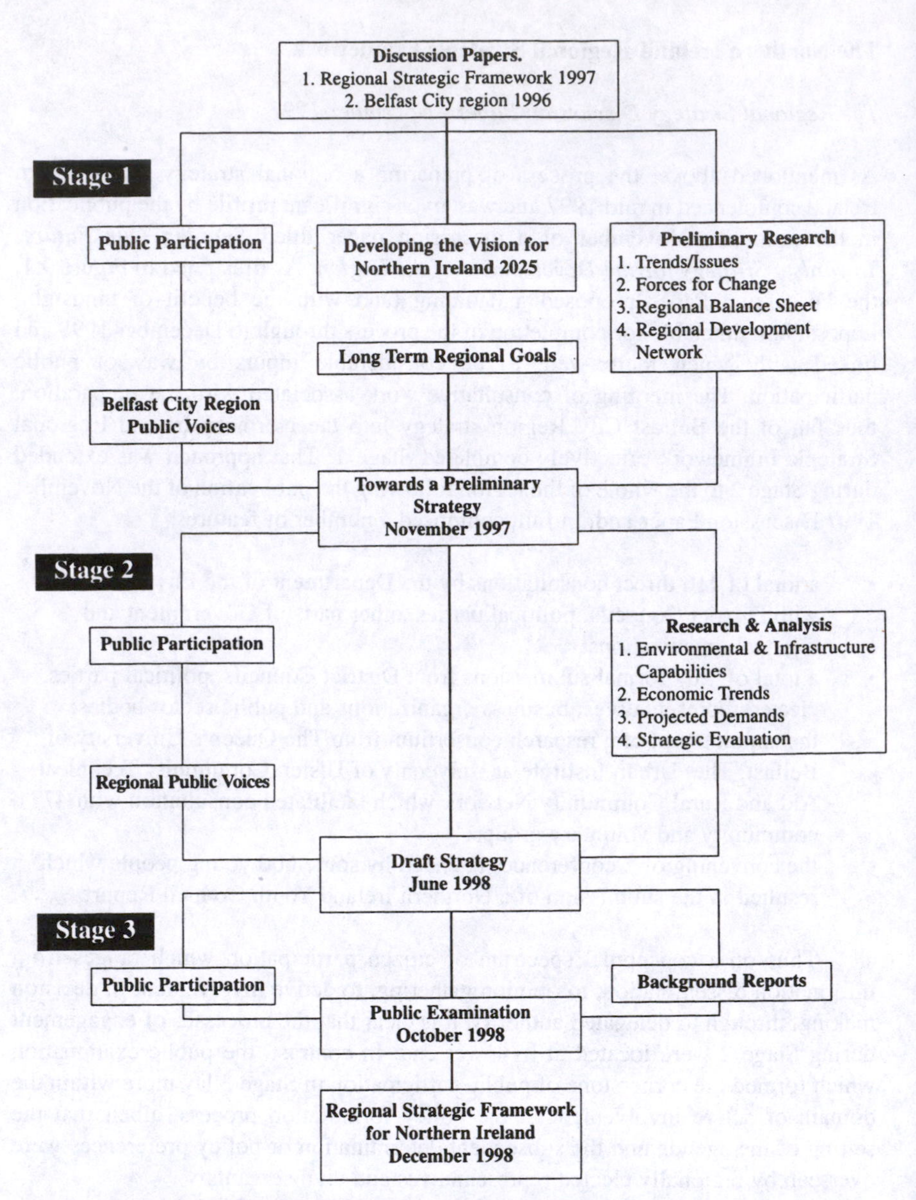

Source: Department of the Environment for Northern Ireland (1997) *Shaping Our Future - A Discussion Paper*. Belfast: Regional Planning Division.

Figure 9.1 The Proposed Northern Ireland Regional Strategic Framework Process

Belfast and the region's towns and rural communities; only Belfast's core evidenced population decline. Particular reference was made to the rapid growth of the rural population in Northern Ireland and the possibility of it becoming an environmental threat was hinted at. Elsewhere the Discussion Paper outlined a strong protectionist theme as part of that environmental agenda - the designation of extensive protected areas, goals to conserve agricultural land, the application of resource, environmental capacity and transport tests to new development proposals, and a more compact rather than dispersed settlement strategy. The cities along with a number of towns across Northern Ireland were identified for high, medium or low growth within a long term development scenario to 2025. The necessary expansion of the tourism sector was linked to the protection and management of the rural environment and the dangers of blurring town and countryside were spelt out.

Not surprisingly, the response to this prescription from rural interests was comprehensively critical (McEldowney *et al* 1998). A major conference convened as part of the consultative exercise attracted over 200 representatives from District Councils, District Partnerships, rural development local action groups and community associations. This was followed up by 25 workshop sessions across a wide territory. Consultees welcomed the preparation of a Northern Ireland strategic framework and the opportunity for dialogue, but expressed concern at its perceived urban bias, the dominance of the Belfast City Region and the hurried timescale for submissions. A common perception across rural Northern Ireland was that the ghost of the 1960s regional plan, with its growth center bias, had been awakened. Instead there was strong advocacy for a living and working countryside, a diverse economy and a settlement hierarchy which included dispersed rural communities. A preferred planning principle was 'valuing people' which was linked with 'social and spatial equity'. The depth and breadth of this criticism did succeed in putting pressure on the regional planning team to rework its preliminary proposals.

The Draft Regional Strategic Framework - December 1998

The draft Regional Strategic Framework (RSF) for Northern Ireland was subsequently published in December 1998. It was not surprising that the content and tone were, at first sight, much more conciliatory to rural interests. For example, there was new recognition of the 'Rural Community' and 'Rural Northern Ireland' which, within a chapter setting out the overall spatial development strategy, were actually discussed ahead of the Belfast Metropolitan Area and Derry/Londonderry as regional cities!

Within the draft RSF the previous designations of high, medium and low growth towns were replaced by a two tier Regional Town hierarchy of Major Service Centers and Key Service Centers (Figure 9.2). Housing targets through to 2010 were ascribed to each District Council area, its Regional Town and the Rural Community component of the District. The identification of key transport corridors was augmented by the inclusion of strategic trunk roads and the different elements making up a regional strategic transport network were given fuller expression. Also of particular significance was the articulation of not fewer than 30 draft strategic planning guidelines (SPG). These ranged widely from the global to the local and from urban to rural, but essentially were written as 'feel good headlines' which could be expected to attract little criticism. For example:

SPG 1: *To strengthen and extend European and world-wide linkages;*
SPG 2: *To increase links with neighboring regions and capitalize on trans regional development opportunities;*
SPG 3: *To foster development which contributes to community relations, recognizes cultural diversity, and reduces socio-economic differentials;*
SPG 6: *To build local communities for the future in existing urban areas;*
SPG 9: *To sustain a living and working countryside;*

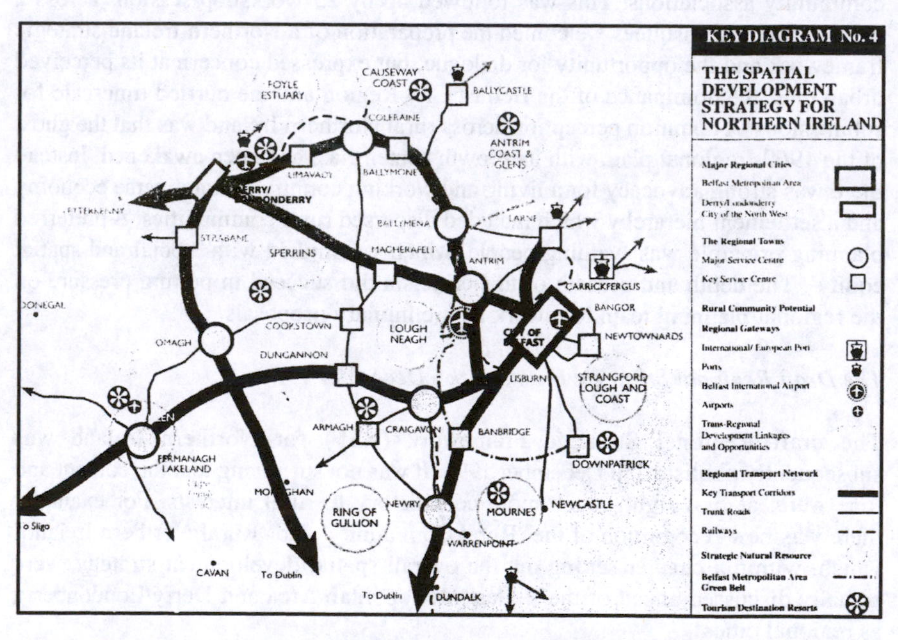

Source: Department of the Environment for Northern Ireland (1998) *Shaping Our Future - Draft Regional Strategic Framework for Northern Ireland.* Belfast: The Stationery Office.

Figure 9.2 The Draft Regional Strategic Framework Spatial Development Strategy

A total of 169 representations in relation to this document were received by the Department of the Environment through to April 1999 from District Councils, political parties, organizations and individuals. Within rural areas the draft RSF was welcomed as a significant advance in thinking compared with the Discussion Document of November 1997. Nevertheless, elected representatives across Northern Ireland voiced reservations in submissions to the Department of the Environment. For example, the differentiation between Major Service Centers and Key Service Centers was perceived as an artificial construct which could arguably harm the prospects of the latter. The explication of SPG 8 - to constrain the future growth of small towns and villages, it was argued, sat uneasily with the thrust of SPG 9 - to help rural communities develop strategies for economic and social regeneration. Local authorities also took the view that the draft Strategy as a whole did not sufficiently respond to the challenge of joined-up government, especially in regard to rural development, education and health care provision and indeed had tended to stay clear of important, though controversial, areas of policy formulation work. Submissions argued for the opportunity to further discuss concerns at the forthcoming public examination.

The public examination into the Regional Strategic Framework

A public examination is designed to provide information through informed discussion of certain matters relevant to policy decision making. It provides an arena within which different stakeholders can present facts and arguments to an independent panel whose task, in turn, is to make recommendations on the basis of the evidence to the sponsoring authority. Within Great Britain(GB) the public examination tool has long been associated with the preparation of Structure Plans and latterly has been adapted to test the content of draft Regional Planning Guidance. But within Northern Ireland, that tradition of policy deliberation has not been a feature of physical planning which has tended to rely on the conventional, highly formal and adversarial public inquiry process to deal with land use and development issues associated with statutory zoning plans. The decision to include a public examination as Stage 3 of the program for the preparation of a Northern Ireland regional strategy can be viewed, therefore, as a welcome innovation in dialogue based participatory planning.

The public examination into the draft RSF was convened over a five weeks period during October and November 1999. Prior to that the appointed panel of three members along with its secretariat had:

- made themselves familiar with the GB public examination experience;
- reviewed all 169 submissions received by the Department of the Environment in relation to the Draft RSF;
- selected and revised, following additional consultation, the matters to be included in the public examination;

- selected the participants;
- invited written statements from participants on the selected matters;
- convened two public meetings to provide procedural information;
- agreed different venues across Northern Ireland to hold the public examination; and
- conducted a pre-meeting on the central issue of housing projections.

Over 170 different participants attended the public examination on behalf of a wide range of invited organizations drawn from central government departments and agencies, District Councils, the business, community and voluntary sectors, and representative bodies. Their input was arranged by the panel to be in line with their previously expressed concerns and expertise, and while a number of organizations appeared at more than one debate, any single debate did not usually have more than 24 participants. The formation of strategic alliances was very much encouraged by the panel and resulted, for example, in several community bodies concerned, *inter alia*, with protection of the Belfast green belt coming together to form the Belfast Metropolitan Residents Group. Similarly, a group of five local authorities in the rural periphery of Northern Ireland decided to press their case collectively under the banner of the West Rural Region, with spokespersons comprising elected representatives and chief executives.

The report of the panel which conducted the public examination into the draft RSF was subsequently published in February 2000. It runs to 130 pages and contains some 110 recommendations. A follow-on response by the Department for Regional Development (DRD), which assumed responsibility for regional planning following devolution in December 1999, appeared in April 2000 and, over 35 pages, comprises a series of statements which indicate acceptance of, a willingness to take note of, or an undertaking to give further consideration to these recommendations.

To sum up at this stage of the chapter, therefore, it is the case that the *Shaping Our Future Discussion Paper* of November 1997 anticipated the publication of a final version of the RSF in December 1998. This expectation was not realized. This is not to argue that what was achieved, has been of little value. Rather, the emergent and uncertain shape of the policy content confirms the complexity of the regional planning challenge which must struggle in balancing inclusiveness during formulation with effectiveness during delivery. What is also quite clear is that the maximizing of effectiveness must deal with inevitable policy segmentation across departmental and agency structures. This has, arguably, been made even more difficult by the fragmentation of the Department of the Environment, which initiated the preparation of the RSF, into three new departments following devolution (Department for Regional Development, Department of Social Development, Department of the Environment) and the development linkages which cross-over to the other eight departments under the new Northern Ireland Executive. The iterative nature of the policy formation-implementation process further constrains the realization of policy content comprehensiveness at any one point in time.

At a conceptual level participation is essentially concerned with involving the people affected by public policy decisions in helping to make, implement and monitor the policies from which those decisions derive. Consultation as an important element of the spectrum of participation has clearly been a major feature of the Northern Ireland regional planning process. But it should also be noted that action research by Rural Community Network (1999) has identified considerable consultation fatigue among participants due to the sheer number and diversity of public policy initiatives brought forward within Northern Ireland in recent years. Contributing factors include the use of boring techniques, the perception that the process is extractive, the absence of an enriching experience, limited advance information sharing, the same people attending meetings and a concern that public officials do not really listen to community inputs. The research has also identified that women, members of the Protestant community, Travelers, the long term unemployed, and farmers are often the most excluded from participation in the range of consultation exercises undertaken during the past decade. In an effort to connect physical planning policy more sensitively with the management of growth and development across Northern Ireland it would seem that the participatory processes connected with the preparation of the Regional Strategic Framework have at least had the potential to address some of these shortcomings by creating a necessary and lengthy space for active involvement, culminating in policy related dialogue through the public examination mechanism. At this juncture it is appropriate, therefore, that the following section in this chapter looks more closely at this high watermark of deliberative participation by reviewing the results of a questionnaire distributed to invitees attending the public examination. Notwithstanding the Northern Ireland focus, the data in their own right should command wider interest from policy makers and planning practitioners with an interest in the public examination process.

The questionnaire survey of public examination participants

An important part of this research into planning through dialogue involved the administering of a questionnaire to all participants attending the public examination into the draft RSF. The aim of the survey was to provide an opportunity for individuals to express their opinions on the conduct of the public examination and the perceived relevance of this tool for other areas of public policy debate in Northern Ireland. The questionnaire was prepared mid way through the public examination hearings and was distributed by post during the week following their completion in order to tap into experiences that were still fresh in respondents' minds. The mailing list was drawn up on the basis of daily attendance lists compiled by the public examination secretariat. The questionnaire sought to obtain views on (1) the debate themes and the related discussion topics; (2) matters connected to preparations for the public examination in relation to the quality of briefing papers, the procedural guidance given by the panel, and the usefulness of the public examination library and Internet web site; (3) the participatory experience at the public examination; and (4) the scope for broader adoption of the public examination process and ways to improve its effectiveness. The following analysis is based on 86 returns out of 166 distributed, representing an overall response rate of 52 per cent. With respect to sectoral representation, Table 9.1 indicates the stratification of the survey population and respondents.

Table 9.1 Survey Population and Respondents by Category (% rounded up)

Category	*Population*		*Respondents*	
	No	%	No*	%*
central government	11	7	4	5
district council	48	29	21	24
other public body	29	17	9	11
private organization	13	8	13	15
community/voluntary sectors	38	23	26	30
professional body	8	5	3	4
consultant	19	12	8	9

* There were 2 additional respondents who did not declare their category

Responses from the public sector and consultants were slightly below their overall representation at the public examination with the largest element in both the survey population and the categories of respondent being district council members and officials. On the other hand, very good response rates were obtained from participants attached to private organizations and the community / voluntary sectors. The response rate from professional bodies was broadly in line with their representation at the public examination. The important point here, however, is that the questionnaire survey was successful in reaching out to the diversity of public examination participants.

There were 14 different debate topics (Table 9.2) identified by the panel for the public examination, each of which was allocated, in advance, a length of time. Thus, for example, "A Coordinated Strategy?" and "The Overall Strategy" were each given a half day, "Rural Northern Ireland" was awarded a full day and the "Housing Distribution" debate took place over two days. The time allocated to each debate could be regarded quite simply as a reflection of the complexity of the subject matter and the number of participants invited to the discussion. The participation of questionnaire respondents across the range of debates, and who in some instances attended more than one session, is set out in Table 9.2.

Table 9.2 The Public Examination Debates and Participation by Respondents

Debate Title	*The involvement of respondents*	
	No	% of returns
A Coordinated Strategy?	22	26
Overall Strategy	19	22
Spatial Development Strategy	22	26
Developing a Modern Integrated Transport System	19	22
Rural Northern Ireland	20	23
The Environment	14	16
Supporting the Growth of Tourism	17	20
The North West	9	11
Housing Projections and the Management of Housing Growth	15	17
Housing Distribution	18	21
Achieving Quality Built Environments	12	14
Retail Development	11	13
Spatial Development Strategy for the BMA and Wider City Region	22	26
Implementation, Monitoring and Concluding Views	26	30

Again the data underline the representativeness of the survey respondents who are spread across all debate topics. Thus, for example, 11per cent, attended the session on the "North West", while 30 per cent contributed to the final debate on "Implementation, Monitoring and Concluding Views". Respondents were initially asked to comment on the themes selected for the public examination debates. Some 44 per cent took the view that these were adequate for the intended purpose, though a minority of respondents (23 per cent) did not regard the themes as being satisfactory.

Each theme listed by the panel was accompanied by a series of questions which were designed to steer the content of the discussion. Thus, for example, the "Rural Northern Ireland" debate was structured around five questions: (1) Does the strategy clearly set out how sustainable development will be achieved in rural areas? (2) Does the strategy define sufficiently clearly what the future components of the rural economy could be? Does the strategy provide adequate proposals to promote a strong and diversified rural economy? (3) Would the spatial development strategy and its transport system improve the accessibility of the rural community to support services or would a more dispersed pattern of settlement and services be likely to meet rural needs? (4) Would the guidelines for the development of local rural centers help to sustain a living and working countryside? (5) Would the strategy's targets for housing growth provide adequately for local housing needs in rural communities in the period up to 2010? Some 35 per cent of respondents commented favorably on the overall framing of these questions, although an almost similar number (31 per cent) did express reservations.

The panel introduced a number of measures in order to enhance familiarity with the public examination process and to deepen understanding by participants of the many issues in the draft RSF to be addressed during the debates. All participants were asked to submit advance written statements in response to the questions set out under each theme. These were published on the Internet and were made available in a special library facility for open consultation along with other core policy documents. The panel also published information on the purpose and format of the public examination and convened two preliminary public meetings in May and September 1999 to clarify procedural matters raised by participants. Again the minutes of both meetings were published. The Department of the Environment in the period leading up to the public examination (and continuing through the debates) issued a series of additional briefing papers which sought to explain the rationale for different aspects of proposed policy and to update population and housing data. For example, a substantial volume of information was made available at the second preliminary public meeting relating to District Council area housing allocations for the period 1996-2010 along with an evaluation framework for the regional towns. A revised version of housing growth estimates, amounting to an increase of 40 per cent, was subsequently produced by the Department of the Environment during the "Housing Distribution" debate and at which, additional adjustments were made to accommodate anticipated overspill from the Belfast Metropolitan Area and second home development in selected Districts. A total of 86 per cent of questionnaire respondents rated the Department of the Environment briefing papers as satisfactory, very good or excellent, with only eight per cent regarding them as poor and a further four per cent as very poor. A follow on question asked participants to comment on the quality of the written submissions prepared by different organizations attending the public examination. Again a very high 91 per cent deemed these to have been satisfactory, good or very good. A minority of five per cent were of the opinion that they were poor or very poor.

The public examination library was initially available for consultation at the secretariat offices in Belfast and was subsequently set up in a resources room with free photocopying facilities adjacent to the debating chamber for each session of the public examination. The library, in effect, moved from venue to venue with the panel and was available to participants and the general public. While 42 per cent of respondents offered no comment in regard to the usefulness of the library, 34 per cent were of the opinion that it was extremely useful or somewhat useful, and 24 per cent suggested that it was of limited use.

The establishment of an Internet web-site to publish all background papers was an innovation in the participatory planning process in Northern Ireland. A surprisingly high 54 per cent of respondents commented that this was extremely or somewhat useful, with only 14 per cent of participants holding the view that this medium was of limited use. A large 33 per cent of respondents chose not to comment one way or the other in response to this question.

The final matter pursued in the questionnaire regarding preparations for the public examination debates was the level of procedural guidance satisfaction among participants. A very high 87 per cent regarded the work of the panel on this front as being extremely or somewhat helpful, with only 11 per cent holding the view that the panel's efforts were of limited use.

The public examination sat for a total of 16 days over a five week period commencing in Belfast, then moving to two regional towns and returning finally to Belfast. The selection of what could be regarded as two outreach venues was highly symbolic, not least in regard to the substantive debates scheduled in these locations. Thus in one instance, within the above mentioned West Rural Region, the sessions were given over to "Rural Northern Ireland", "The Environment" and "Supporting the Growth of Tourism". A total of 78 per cent of questionnaire respondents expressed satisfaction or extreme satisfaction with the selection of venues, while 18 per cent were dissatisfied or very dissatisfied.

A key difference between a conventional planning public inquiry and the Draft RSF public examination is that entry to the latter was by invitation rather than of right. In its published report (Elliott *et al*, 2000) the panel expressed its recognition of the need for inclusiveness and fairness in the selection of participants in order to achieve public confidence in the process. Thus while it was not possible to offer a seat to every participant at every session, it was still necessary to ensure that those most relevant to a discussion were accommodated. Participants were invited on the basis of their anticipated expertise and breadth of representativeness. The Department of the Environment attended every session and was supported, as the panel felt appropriate, by other public bodies. At each debate, the panel provided a single speaking point for each participant organization in a square table format and, in its own words, sought to discourage "the reading out of formal statements, lengthy lectures or repetitions of views already presented"(p20).

The panel regarded it as its duty to "inquire into and challenge the points made" by participants, and where there was a challenge to the views put by others during any debate, the panel was "content to accommodate questions put through the Chair". The panel noted in its report that none of the participants used the services of legal counsel and indeed restated the view that formal cross-examination of participants should not be a part of the public examination process. It is against this context that participants were asked in the questionnaire to reflect on the overall quality of their experience at the public examination. It was important, therefore, that the questionnaire was distributed immediately following the closing of the public examination when impressions were not dulled by the passage of time. On a scale of 1 to 5, where 1 corresponded with "strongly disagree" and 5 with "strongly agree", respondents were asked to express an opinion in regard to a series of propositions. These statements and their mean scores are set out below:

- I did not feel intimidated by the occasion 3.9
- I was accorded sufficient opportunity to make my case 3.9
- I was able to interactively debate issues with other participants 2.6
- I was happy with the answers which I received to my questions 2.7
- The panel contributed substantially to the debates through its questions 3.2
- I was able to understand the arguments being made by others 4.1
- The content of the debates was coherent 3.6
- I was impressed with the overall quality of the submissions 3.5
- My participation at the public examination has been worthwhile 4.0

While most of the statements attracted reasonably strong support, it is perhaps interesting to note that respondents were less content with the opportunity provided to interact with other participants (52 per cent disagreed or strongly disagreed with the statement) and with the answers received to questions (41 per cent disagreed or strongly disagreed with the statement). Overall, however, 70 per cent of participants agreed or strongly agreed that their participation had been worthwhile, with only six per cent disagreeing or strongly disagreeing with that proposition. In regard to the mix of participants at the various sessions, 68 per cent of respondents were of the opinion that this was excellent or just about right, though a sizeable number of respondents (27 per cent) commented that the mix was not that good or very unsatisfactory.

As outlined in Chapter One, public participation in the policy formulation arena is now part of the mainstream. Within the sphere of town and country planning there are opportunities for public input to the decision-making process in regard to the determination of individual planning applications and in regard to the preparation of development plans. While both development control (on occasions), and forward planning (usually in regard to zoning plans), make use of public inquiries to facilitate that input, the ultimate purpose of the public inquiry remains similar to a public examination. Each is concerned with gathering and testing information and giving best advice in the form of recommendations to decision makers. Two essential differences, however, are rights of access to and the formality of the public inquiry process. This tends to adopt a quasi-legal ritual of witnesses giving evidence, followed by cross-examination and closing submissions. It is this distinct contrast in style which prompted the inclusion of a question on the suitability of the public examination process for testing zoning plans in the future. Some 64 per cent of respondents expressed views, often of conditional support, in favor of using the public examination process in this manner, with 25 per cent opposed to any replacement of the public inquiry tool largely because of the need to ensure maximum representation of individual land ownership rights.

Within Northern Ireland, Government in recent years has published a raft of policy documents relating to health, economic development and the draw-down of future EU structural funds. Devolution has created further potential for fresh and ongoing policy prescription. Accordingly, participants, on the basis of their experience with the Regional Strategic Framework, were asked to comment on the extension of the public examination process to test other Government strategies and programs. A very high 87 per cent of respondents expressed opinions in favor of adopting this approach more broadly, with only ten per cent commenting to the contrary.

In short, the evidence, viewed in the round, indicates that the public examination was a valued part of the participatory process associated with the preparation of a regional strategic framework for Northern Ireland. The majority of respondents regarded the themes selected for the 14 debates as being adequate for their intended purpose, though there is a consistent body of criticism that cross linkages with other areas of public policy were insufficiently developed. This ties in with other observations that the participation of public bodies, outwith the Department of the Environment, was weak. They did not contribute fully to the publication of briefing papers and to the debates. Overall, the contribution of the Department of the Environment in facilitating the public examination, in preparing briefing papers, and in offering comment on various issues was well received by participants. The work of the panel in designing debate questions in advance was appreciated as a contribution to the thinking of participants, though opinion varies across respondents as to whether these were interpreted too flexibly or imposed a constraint on the discussion. Finding that balance is obviously difficult and much rests, therefore, on participants being able to respond to appropriate steers from the panel during each debate. In this regard the interventions by the panel in the form of additional questions were a highly valued part of the public examination. The majority of respondents agreed that their participation at the public examination had been worthwhile, that they did not feel intimidated by the occasion and that they were afforded sufficient opportunity to state their case. However, a significant number of respondents were less content with the scope offered for direct interaction with other participants, and with the answers received to questions when these were passed on through the Chair. A strong preference runs through the comments for more scope to challenge and discuss rather than have people rely mainly on prepared statements. There is substantial conditional support for the adoption of the public examination tool as part of the statutory development plan preparation process, though it was wisely pointed out by respondents that the rights of individuals affected by land zoning would have to be safeguarded through a conventional public inquiry. There may, therefore, be scope for introducing both procedures into the preparation of future development plans. And finally, participants were very enthusiastic that other Government draft strategies and programs should be subject to some similar form of public examination.

Conclusion: the implications of the Northern Ireland experience for participatory planning

Participatory planning as dialogue

The preparation of the Northern Ireland regional development strategy has involved an intensive four years discourse of challenging and reframing policy perspectives. It has not been a quick fix, tokenistic exercise. The process as a whole has been highly inclusive and stands in marked contrast to previous expert dependent and centrist technocracy regional planning prescriptions. *Shaping Our Future*, rather, carries the hallmarks of social negotiation as, over time, effort has been directed at moving appreciation of the content from strident adversarialism to quiet consensus. Information dissemination through conferences, the sharing of substantial written material, and consultative workshops on a cross-sectoral basis, culminating in the participation of key stakeholders at a public examination, have all assisted with that endeavor. There is evidence, as Healey (1997) identifies more broadly, that this activity has allowed participants to learn from each other - on what they care about and why - through a combination of rational technical arguments, anecdotal observations, emotional responses and moral advocacy in multiple arenas both informal and formal. Thus, for example, the public examination highlighted some enduring tensions between 'fact' and 'value' as evidenced by the debate between Belfast as the longstanding economic engine of the region and the often expressed preference, based on spatial equity, for greater dispersal of growth opportunities. The totality of this dialogue, however, at the very least, enhances the legitimacy of the planning process and points towards a new way of doing the business of public policy formation. The key issue to emerge, with wider implications, is the extent to which Government is willing to commit itself to participatory permanence which could go as far as requiring each department and agency to have a participation or involvement strategy. A clear message coming through from public examination participants, for example, is greater use of this instrument by the public sector as a whole to test draft policies through conversation, as between their consistent application across Northern Ireland and their responsiveness to local situations.

Participatory planning and democratic renewal

The educational and transformational relationship between participatory processes and democratic renewal is important. The renewal of trust between central government and local government, as well as harnessing the latent energy of citizens to build a vibrant associational sector are core objectives in the fostering of that linkage. Participatory approaches by themselves will definitely not create democratic renewal. A broader debate is required on that front about the value and role of local democracy. However, participation must be regarded as an important part of that activity and, as argued in Chapter One, must be associated with the development of a good mix of consultative and deliberative methods. Participation should be regarded as an ongoing activity and it must bring added value to policy formulation and implementation by way of greater responsiveness and ongoing interactive learning. Moreover, it would seem important that the impact of participatory processes is carefully evaluated and that the results of this work are made widely available to enhance this iterative learning. Ultimately what is being strived for here is the mediation of differences through dialogue.

Within Northern Ireland it could be argued that the *Shaping Our Future* process has contributed to a measure of democratic renewal in a society where local political debate has for so long been less concerned with environmental, social and economic concerns and has been constrained by a combination of bureaucratic hegemony and Direct Rule whereby Government Ministers are not regarded as being directly accountable to the regional electorate. The preparation of the Regional Strategic Framework has clearly captured the involvement of an active citizenry across a very wide spectrum of issues. It has generated community coalitions, has provided a stimulus for the creation of new alliances at local government level and has fostered new collaborative relationships between elected representatives and the constituents that they seek to serve. In short, *Shaping Our Future*, has helped provoke a fresh enthusiasm in citizen planning at the strategic level when so rarely any involvement in planning matters rises above entrenched parochialism and territorial dispute.

This chapter commenced by expressing surprise that a participatory planning approach could manifest itself in a society so deeply divided as Northern Ireland. But after over three decades of turmoil the region has faced the prospect of peace following on from the constitutional settlement set out in the multi lateral Good Friday Agreement of April 1998. Any successful peace process depends on many inter related factors comprising the need to deal with a culture of mistrust and violence, moderate the attachment to historic symbols and ritual, search for constitutional and institutional consensus, and secure regional stability and prosperity. A regional planning process as full as that engaged in within Northern Ireland has the capacity to strengthen a broadly defined civil society as an expression of the popular desire for peace. Participatory planning through sustained dialogue can help to create that shared vision of a better and a different future.

Acknowledgment

The authors acknowledge grant support for the research presented in this chapter from the Rural Innovation and Research Partnership (NI) Ltd, funded through the EU LEADER 2 program. An extended version of this chapter was first published in *Policy Studies*, Vol 23, No 3 / 4, 2002, pp191-209.

References

Department for Regional Development (2000) *Shaping Our Future – The Regional Development Strategy for Northern Ireland: Response by the Department for Regional Development to the Report of the Independent Panel following the Public Examination*, Belfast.

Department for Regional Development (2001) *Shaping Our Future – Regional Development Strategy for Northern Ireland 2025*, Belfast: Corporate Document Services.

Department of the Environment for Northern Ireland (1997) *Shaping Our Future – A Discussion Paper*, Belfast: Regional Planning Division.

Department of the Environment for Northern Ireland (1998) *Shaping Our Future – Draft Regional Strategic Framework for Northern Ireland*, Belfast: The Stationery Office.

Elliott, A., Forbes, J. and Robins, D. (2000) *Shaping Our Future – Draft Regional Strategic Framework for Northern Ireland: Report of the Panel Conducting the Public Examination*, Belfast: Department for Regional Development.

Healey, P. (1997) *Collaborative planning: shaping places in fragmented societies*, London: Macmillan Press Ltd.

McEldowney, M., Sterrett, K., Gaffikin, F. and Morrisey, M. (1998) 'Strategic planning in Northern Ireland: some reflections on contestation and consensus', *Pleanail – The Journal of the Irish Planning Institute,* Vol 14, pp111-116.

Rural Community Network (1999) *Policy change and conflict resolution in rural areas – report on feasibility*, Cookstown: Rural Community Network.

Chapter 10

From Enemies, to Higher Ground, to Allies: The Unlikely Partnership Between the Tobacco Farm and Public Health Communities in the United States

Frank Dukes

> At first people refuse to believe that a strange new thing can be done, then they begin to hope it can be done, then they see it can be done – then it is done and all the world wonders why it was not done centuries ago.
> Frances Hodgson Burnett. (A Secret Garden)

> When people of good will get together and speak from their head and heart, and share their ideas for the future, good things happen. (Tobacco grower)

> We've learned that you have to do something different to get something different. (Health advocate)

> The two groups have proved that a coalition is stronger than the two parts acting separately; we can walk side by side to promote public health. (Tobacco grower)

> We have the potential to become something even more powerful … literally changing the course of history. (Health advocate)

Introduction

Tobacco farming and public health concerns may seem entire incompatible. Indeed, for many years the relationship between tobacco farming and public health communities in the United States could best be described as both distant and antagonistic. In the mid-1990s American farmers stormed tobacco state' capitals and Washington, D.C. with armies of tractors to protest the possibility of Federal Drug Administration (FDA) regulation of tobacco products, which was (and remains) the tobacco control advocates' highest priority. At the same time, health-friendly legislators in Congress periodically introduced legislation to weaken or dismantle the federally-administered tobacco program,[2] the maintenance of which was (and remains) the highest priority of tobacco farmers. As late as Spring, 1997, a prominent public health advocate described the nascent relationship between tobacco growers and public health advocates by asserting 'There is no common ground'.[3] She was correct in one sense, in that common ground is not to be found just by looking; like the farmland that produces the golden leaf, common ground needs the hard work of cultivation. That she was incorrect in the most important sense has been demonstrated many times since then:

- On March 18, 1998, when a press conference featuring three tobacco grower leaders and four public health leaders announced the 'Core Principles Statement between the Public Health Community and the Tobacco Producers Community'. Since that time over 70 organizations and many prominent individuals endorsed those Core Principles, including major tobacco farm and public health organizations and individuals ranging from then-President Bill Clinton to former President Jimmy Carter to evangelist Pat Robertson.
- On March 30, 1999, when Virginia's Governor Jim Gilmore signed historic legislation that sent 50 per cent of Virginia's share of the tobacco companies' settlement with states' attorneys-general (which will total an estimated $4.2 billion) to indemnify tobacco farmers for lost income and to promote economic development in tobacco dependent communities, and 10 per cent of that share to aid tobacco control efforts for Virginia's youth.
- On May 14, 2001, when a Presidential Commission made up of tobacco farm and public health leaders issued a series of joint recommendations that included FDA authority over tobacco products, an increase in the federal cigarette tax of 17 cents, a tobacco quota buyout, and development of a radically revised tobacco farm program.

These agreements and other related actions did not arise because the two sides wanted to be nice to one another. Nor did they merely reflect a relatively common temporary compromise between continuing adversaries. Rather, these agreements represent a fundamental realignment – an improbably strong alliance, developed from deliberate efforts to foster constructive dialogue and to create productive relationships based on mutual interests, integrity, and respect.

As a mediator who has helped convene and facilitate many farm-health discussions through the Virginia and Southern Tobacco Communities Projects, and who has watched mutual suspicion and antagonism change to shared trust and cooperation, I would like to describe in this chapter the history of what one critichas called an 'unholy alliance', the benefits these new relationships have provided to participants, and the prospects for any lasting alliance between the two groups.

An uncertain future for tobacco farming and public health

The 1990s began a period of hard times for U.S. tobacco producers and tobacco producing communities, with an economic decline that continues as this book goes to print. For a variety of reasons primarily involving tobacco company strategies and improved production from tobacco farmers around the world, the amount of American tobacco produced and sold has declined sharply. Most analysts foresee continued decline for some years before some period of stability on the farm eventually returns. That decline translates into real hardship for tens of thousands of farmers and their families and hundreds of farm communities dependent on tobacco leaf income, including affiliated business such as tobacco warehouses, processing plants, and farm equipment retailers.

At the same time, tobacco use continued to extract a heavy toll on health worldwide. Despite complaints of heavy-handed government regulation by the tobacco companies, tobacco usage around the world actually continued to grow as new markets were being opened and new products were being introduced. Health experts predict substantial growth in tobacco-related illness in the foreseeable future. As of 2002, an estimated 400,000 people in the U.S. and 10 million people globally die each year from tobacco-induced illness.

The tobacco farm community – public health community relationship

Until 1994 these two conditions – the economic peril to tobacco farming communities, and the disease and death caused by tobacco use – were not considered related by farmers or public health advocates. When farmers and farm leaders spoke of public health issues, the tone and content was defensive and hostile. A tobacco farm leader might reason that the public health advocates were the 'anti's' who wanted to ban tobacco products; their concerns were exaggerated or nonexistent; tobacco was being singled out unfairly for special attack; farmers were unfairly labeled 'drug pushers' for producing a legal product that had provided for their families and communities for generations; none of these 'anti's' ever took time to understand the plight of rural communities; and survival of their farms, their families, and their communities justified any and all legal and legislative strategies.

For many public health advocates the converse was true, albeit with some difference in intensity and focus. These advocates might suggest that farmers were not the target; the tobacco companies were the enemy. But if the farmers got hurt in the way, well, whose fault was that? The tobacco companies lie; their products are lethal and addictive; they are ruthless in their efforts to sell their product around the world; they target children and youth; if sales go down, farmers can just grow something else; farmers and farm legislators are the companies' allies; and preventing the awful toll of disease and death caused by tobacco use justified any and all legal and legislative strategies.

The respective beliefs and attitudes of these two sides were formed almost entirely in isolation from one another. In fact, what each side believed they knew of the other was often determined through deceptive filters and lenses provided by the tobacco companies. The companies, with their powerful lobbyists in state capitals and Washington, D.C., kept a close watch on tobacco control strategies and initiatives. The farmers learned of proposed legislation and regulations through intensive tobacco company communication efforts, including editorials and articles in farm magazines, direct mailings, trade shows and conventions, leadership programs, and special meetings and conference calls convened when the companies determined farmer support was needed.

One example of the nature of such communication and the strength of the company-farmer alliance occurred in 1993. The United States Food and Drug Administration (FDA), under Commissioner David Kessler, had begun to investigate tobacco company practices and concluded that the FDA ought to be asserting regulatory authority over tobacco products. The companies convinced many farm leaders that any such regulation would lead to a significant decline in sales, hordes of federal regulators tramping around tobacco fields, and even complete prohibition of tobacco products. Massive protests involving hundreds of tractors and farmers in state capitals and Washington helped mobilize congressional action that withdrew support for FDA regulation.

Public health advocates in tobacco-producing states and at the national level certainly heard much about the farmers, albeit indirectly. Tobacco-control efforts – for example, to provide public funding for tobacco cessation programs, or to raise state tobacco excise taxes, or to allow localities to determine tobacco use policies rather than have such policy dictated at the state level – were met with a constant refrain from tobacco state legislators of 'I'm not doing anything that will hurt the tobacco farmers'.

First steps – Virginia Tobacco Communities Project (VTCP)

In the early 1990s, a small group of public health advocates frustrated with their inability to make even incremental improvement in reducing tobacco-related disease determined that an entirely new approach was warranted. As Rebecca Reeve, one of the leaders of this effort says, 'If you always do what you've always done, you will always get what you've always got'.[4] They decided to initiate a process of contact, dialogue, and mutual education with tobacco producing interests.

This initial contact was not entirely unprecedented; Jimmy Carter had convened a one-time meeting with tobacco farm leaders and leaders from the public health and medical communities. However, any relationships initiated during that meeting were not sustained. By the early 1990s the idea that there could be any benefits to discussions among the two groups was held by no more than a handful of individuals. Indeed, many people in both camps thought that irreconcilable differences between the two groups left nothing to talk about. For many, if not most, even to talk with someone from the other side was equivalent to treason.

This, then, was the climate in which initial engagement began: almost no prior discussions or relationships between the two groups, high stakes and continuing controversy surrounding both tobacco production and anti-smoking efforts, and active opposition to such talks by many members of both groups and the most powerful player in that whole arena, the tobacco companies. So it is not surprising that the effort began with caution and even suspicion, or that progress was slow and uneven.

The initial task was to acquire funding and to put together a project team which would lead this initiative. Rebecca Reeve of the Institute for Quality Health (IQ Health), part of the University of Virginia (U. Va.) Health Services Foundation, led a team of health advocates that also included the Virginia Chapter of the American Cancer Society and, to a lesser degree, the Virginia Department of Health. Recognizing that tobacco-growing interests would be skeptical about the motivations of health advocates, IQ Health sought the services of professional, independent facilitators from U. Va.'s Institute for Environmental Negotiation (IEN), with myself as lead. An economist from U. Va., John Knapp, was contracted to provide an economic analysis of the tobacco industry in Virginia. Funding was solicited and received from the Robert Wood Johnson Foundation, and the Virginia Tobacco Communities Project (VTCP) was born.

Entry into the Virginia tobacco farm community

Once funding was acquired in August of 1994, the next step was to gain entry into the Virginia tobacco-growing community of farmers, agriculture extension agents, university researchers, government entities and associated organizations. Independently, a Joint Legislative Study Committee on Alternative Strategies for Assisting Tobacco Farmers, chaired by the state delegate representing Charlottesville, the home of the University of Virginia, was also convened at that time. The existence of this state legislative committee provided both credibility to VTCP efforts and an influential receptor for potential project recommendations. Members of the project leadership team attended an early two-day session of this Committee involving visits to a tobacco farm, a tobacco warehouse and auction, and a tobacco processing plant. These visits provided informal opportunities to discuss the VTCP with potential participants and to make the personal contacts necessary to get the project off the ground.

Virginia tobacco 'Roundtable' meetings

The initial plan of the project team was to convene an ongoing group of diverse interests who would attend a series of "Roundtable" meetings about economic diversification in tobacco-growing communities. The Roundtables were expected to react to and critique analyses of the current and developing changes in tobacco communities. As experts on their own situations, members of the Roundtables could provide community-based strategies grounded in the reality of tobacco communities.

As lead mediator/facilitator, I spent many hours calling several dozens of potential participants and explaining the goals and process of these Roundtable meetings. The initial meeting, held in Danville, Virginia on November, 1994, featured some 40 individuals discussing the prospects for tobacco farming, the need for and interest in agricultural diversification, and economic development in general. The purpose of this meeting was to introduce the project to the tobacco farming community in Virginia and ascertain if there was any support for continuing dialogue. The discussion was awkward: many of the farmers, men and women, looked as though they would rather be anywhere else but there; the few health advocates in attendance hardly said a word; no one spoke of any insights or 'aha's' or even admitted to any learnings at all. But nobody walked out, either, and many of the farmers acknowledged the need to 'do something' to change the situation they were in.

Building upon the telephone discussions, personal contacts, and the limited learnings of the first meeting, a second meeting was convened in March, 1995 with the following stated goals:

- Assess the support for economic diversification in tobacco-growing communities;
- Identify grower and community needs for agricultural diversification and economic development;
- Identify and increase individual and institutional resources;
- Develop a network for community leadership on economic diversification in tobacco-growing communities.

In the meantime, the project team realized the importance of bringing into project leadership the state agricultural university that worked in the tobacco producing regions. An agricultural economist from Virginia Tech University, Wayne Purcell, who had been advocating preparation for change in tobacco-growing areas was willing to participate, and a small amount of funding was secured to assist his participation.

The second Roundtable meeting featured presentations by the two project team economists to about 25 participants. They spoke of a continuing decline in the role of tobacco in the state economy, the potential for even greater decline through a combination of market pressures, taxes, and health concerns, and the difficulty and expense involved in facilitating the inevitable change so that tobacco-growing communities were not devastated. As was the case in the first meeting, the discussion between tobacco farm and public health interests was unremarkable.

A third Roundtable meeting, in April 1995, featured presentations from a sustainable agriculture advocate and former tobacco grower in Kentucky and yet another economist, an individual who had been involved in helping shape tobacco production and marketing policy for decades. This was the most contentious and, not coincidentally, the most productive of all the meetings to date. The elected Virginia representative to the Flue-Cured Stabilization Corporation attended and he and other tobacco farmers revealed the tensions within their own community. The growers spoke candidly of their beliefs that the manufacturers don't care about them, that growers are more valuable to manufacturers as voters than as producers, and that manufacturers give more support to growers overseas than to American farmers. Participants agreed to the creation of smaller working groups to tackle four areas of interest that had been identified during these meetings: 1) continued marketing and production support for tobacco; 2) access to information about profitable supplemental on-farm enterprises; 3) financing for small business development; and 4) education for employment in specific work sectors.

The first three roundtable meetings were held in the flue-cured region of 'Southside' Virginia. A fourth Roundtable meeting, later in April 1995, was held in the burley-growing area of Virginia. Burley growers in Virginia farm much smaller amounts on average of tobacco than do growers of flue-cured tobacco, and they are far fewer in numbers than their flue-cured tobacco counterparts. This meeting was intended to ascertain the concerns and interests of burley-growing communities and to suggest the need for and potential of supplemental enterprises. Presentations by a North Carolina state agricultural development official and Kentucky Community Farm Alliance leaders focused on agricultural diversification. Participants from the burley-growing areas also endorsed the goals of the proposed four working groups.

In June, 1995, the project leaders visited the Flue-Cured Tobacco Stabilization Corporation, the flue-cured tobacco co-operative, in Raleigh, North Carolina. The goals of this meeting with the Board executive director and Virginia's Board member were to explore potential interest in uses of the co-op for non-tobacco production and to plant the seeds for future collaboration and develop personal relationships with Board leadership. As expected, no agreement was pursued; indeed, in one candid exchange a Corporation leader described the discussion in terms of 'holding a snake in a bag'. In June, project leaders also presented their findings and working group plans to the legislative study committee concerned with tobacco alternatives.

Led by the Virginia Tech economist, the four working groups developed their recommendations during the summer and fall. On December 13, 1995, in Roanoke, project leaders made a formal presentation of these recommendations to the Joint Legislative Study Committee on Alternative Strategies for Tobacco Farmers. This presentation was made in conjunction with a dinner and town meeting the evening before.

A town meeting and legislative recommendations

Despite the weather threatening snow and ice, almost sixty people attended the town meeting, including three members of the Legislative Subcommittee and media representatives from as far away as Atlanta. The Roanoke, Virginia public radio station broadcast portions of the meeting in January 1996.

A team of facilitators from the Institute for Environmental Negotiation (IEN) introduced the meeting by providing information about the Virginia Tobacco

Communities Project, the goals of the working groups, and the meeting agenda. Wayne Purcell from Virginia Tech presented a summary of seven recommendations proposed by the study groups. In his presentation he emphasized that what is at stake is not a crop, but people. To make his point he posed the following tobacco family scenarios:

Family 1: Jane and Jack Smith. Full-time farming, 60 years old (just five years older than the average age of farm operators in the top 12 tobacco producing counties in Virginia, where four out of ten farm operators are over 60 years of age), two grown children working in Richmond and Lynchburg. They have 22 acres in tobacco, leasing eight of those, and get half of their farm income from tobacco (in the top 12 tobacco-growing counties in Virginia, tobacco accounts for about half of the dollar value of all farm products sold). They would like to sell their farm and retire in five years.

Needs: To continue to get value out of tobacco production; to get value out of existing quota owned; to get value out of farm when sold or developed.

Family 2: George and Martha Taylor. 42 years old. George is employed by a building supplies firm, Martha works at a tobacco processing plant. They have very little money set aside for retirement. Martha dropped out of high school in the tenth grade, her work is seasonal, and she is concerned that there may be layoffs in the next few years. George has a high school diploma, does household repair work in his spare time, and would like one day to begin his own construction and remodeling business.

Needs: Adult education to get GED; training for skills such as plumbing, construction, or electrical work; small business financing.

Family 3: Sally and Tom Smith. 35 years old. Sally is a full-time homemaker. The Smiths have three children aged 10, 8, and 3. Tom works in retail sales at Southern States. They both help Tom's parents work their farm, which has five acres of tobacco plus a cattle operation. The Smiths would like their children to grow up on the farm. Ideally, they would like to be able to purchase the farm from his parents and farm full-time. They wonder whether they can get a loan to make the purchase, how they will maintain the productivity and profitability of the farm, and whether and how they can purchase new equipment.

Needs: Financing for purchase and capital improvements; advice and expertise for on-farm enterprises; possible short-term and/or part-time employment for one or both to supplement their farm income, at least initially.

Family 4: Jake and Sandra Thomas. 21 years old. Jake and Sandra are seniors at Virginia Tech, the Virginia land grant university. They were married last summer. When they left their hometown of Dungannon, they were not sure they ever wanted to return. They knew they didn't want to spend any more summers working their families' small tobacco plots and the coal companies weren't hiring. Both have traveled around the East Coast for the last three years and they have gained a better appreciation for the beauty of their heritage and the benefits of close-knit family and community. They would like to return somewhere in Southwest Virginia to begin their own family.

Needs: Economic opportunity; sense of possibility, knowledge of existing economic development efforts; high skill or high tech employment opportunities.

The seven recommendations addressed the needs illustrated by the sample families. These needs were centered around four key areas corresponding to the four working groups:

- improvements in production and marketing of tobacco;
- access to information about profitable supplemental on-farm enterprises;
- financing for small business development;
- education for employment in specific work sectors.

Dr. Purcell noted that Virginia farmers are proud people, with a strong work ethic and entrepreneurial spirit. These assets can serve as catalysts for new farming and other money-generating activities. There are three essential ingredients for product success: market power; an open mind in approaching economic adjustments; and community support, including financial assistance. In order for tobacco growers to compete in an environment of changing market conditions, they need information and programs to expedite new business ventures.

The seven recommendations were offered as specific requests to Virginia's Legislative Subcommittee examining the prospects for tobacco farmers. These included requests to restore extension and research funding, to examine the potential of an agricultural Capital Access Program, to endorse a conference on supplemental enterprises and alternative uses of tobacco, to endorse a survey of growers' needs and interests, and to support cooperative efforts for education and economic development in rural communities.

The recommendations were largely endorsed by participants at the Town Meeting. Many tobacco growers continued to push for improvements in production and marketing of tobacco, given the importance of the crop to both the local and state economies. Participants acknowledged the region's dependence on tobacco and its importance to the well-being of its people and the overall economic security of their communities. However, some participants from tobacco-producing communities pointed out that if there are fewer producers in the future due to increased mechanization and economies of scale, there may not be any alternative except to find profitable supplemental crops and business ventures. They acknowledged that growers are not adverse to change if it will bring them a profit. They urged agencies involved in economic development to cooperate to find quality jobs for the younger generation who do not have the capital to buy land and equipment and replace retiring farmers.

Lessons from the Virginia Tobacco Communities Project

There were a number of lessons learned from the experience of the Virginia Tobacco Communities Project. Key substantive lessons included:

- The first interest among growers overwhelmingly is to continue to produce and make a living from tobacco.
- Production, marketing and profitability of tobacco has gone through many changes over the past several decades; however, barring unforeseen and extreme changes, tobacco will continue to be important in tobacco growing states for a long time.
- Nonetheless, there has been a steady, long-term decline in the overall income received by growers and an even greater decrease in the numbers of growers.
- Much of that decline has occurred through increasing international competition, some of which has been supported by American manufacturers.
- There will be continued and even greater concerns about health impacts of tobacco use and continued and even greater efforts to limit its use, particularly among youth.
- There is a need for attention to the needs of tobacco growing communities as they confront changing economic conditions.
- There is considerable, consistent public support for investing public and private resources in these communities.

The profitability of tobacco for growers, a product of limited production (quotas) and artificially high prices (price support), makes for considerable dependence on the crop in tobacco-growing communities. This dependence, fortified by cultural, historical and symbolic representations, is used effectively by tobacco companies to garner sympathy and support for the industry. Health advocates – often termed "anti-s" – are used by the industry as scapegoats for uncertainty and periodic decline in production and/or prices, and, indeed, flat or declining cigarette use and high cigarette taxes are undoubtedly to considerable degree a result of such advocacy. In fact, however, competition from around the world is a substantial and probably primary factor for current pressures to lower prices and profits. Many growers recognize that factor and the role that the big tobacco companies play in supporting growers in other countries, including provision in some cases of seed, fertilizer, technical assistance, and so forth.

Despite considerable antagonism among the growers towards the manufacturers for their support of overseas production and other reasons, their alliance is not likely to be shaken any time soon. This is due to a 1994 tobacco leaf buyout agreement, in which most of the major manufacturers agreed to purchase considerable surplus tobacco stock from previous years. This buyout, which followed a complicated format for discounts of these purchases, meant that growers who had been facing as much as a 40 per cent quota cut for the following year instead were able to increase production. However, this agreement was contingent upon a number of factors favorable to the manufacturers, including the absence of FDA regulation or increased cigarette taxes, a provision which ensured grower opposition to any such regulation or taxes.

The importance of that shift following the buyout, from an expected 40 per cent cut to an increase in production, cannot be overemphasized. The amount of interest among growers in pursuing agricultural diversification and economic development declined considerably.

A project survey conducted by an outside polling firm demonstrated consistent strong support for using a cigarette tax increase to help farmers and farming communities make a transition away from dependence upon tobacco: 74 per cent in tobacco manufacturing areas, 72 per cent in tobacco producing areas, and 76 per cent in Virginia as a whole. Municipalities in Virginia outside of tobacco-growing regions make millions of dollars every year from local tobacco product taxes. Virginia's $02.5 cigarette tax is the lowest in the nation. In private, I had heard a number of people involved in tobacco production say that such a tax would be a good idea. Despite this demonstrated interest in tobacco taxes, strong resistance still existed publicly among producers. In addition to the buyout provisions, farmers are concerned that 1) it would be impossible to keep other interests from demanding a portion of funds raised by such a tax, or 2) a small increase that would have minimal impact on sales might be followed by even greater tax increases and eventual rise in the cost of tobacco products, thus diminishing the demand for their crop.

The current tobacco program limits production through quotas and keeps prices high through the price-support system. An end to that program would likely shift tobacco production further south, increase production, and lower prices. The potential harm to the communities currently dependent upon tobacco production, and the potential impact of increased production and lower prices on the price of tobacco products, raises an interesting question about whether anti-tobacco interests should support continuation of the current program of quotas and price supports.

The main strength of the tobacco program that has allowed growers a good return on their investment is the cooperative arrangement of limited production and price supports. There may be other ways of using existing producer cooperative infrastructure to support non-tobacco economic ventures. This idea was explored with leading tobacco grower interests, but little headway was made. Nonetheless, a project using the burley growers cooperative in Kentucky to market other agricultural products bears further watching. Involvement of large buyers, such as colleges or prisons, may be the key to success in the development of supplements.

The Southern Tobacco Communities Project

Some of these relationships and understandings built during the VTCP may have affected one piece of legislation the following year, as funding for Cooperative Extension agents – a key recommendation of the VTCP working groups – was increased. But as the project wound down at the start of 1996, other recommendations languished. One reason that was evident to VTCP leaders was that many of the problems facing tobacco communities and tobacco control advocates could not be addressed within Virginia, or any other single state.

The meetings that had been held under the auspices of the Virginia Tobacco Communities Project, while not bringing about any major changes or agreements, had at least let people see that there were people to talk to and issues worth talking about. Soon after the VTCP meetings ended, Carter Steger of the American Cancer Society led a team of tobacco control advocates who organized a three-day conference in fall of 1996 called 'Tobacco and Health: Both Sides of the Coin'. That conference, which attracted over 100 public health advocates and a small number of farmers, demonstrated the public health community's interest in the farming issues in a very public way, while educating a large group of health advocates about those issues. Many national and regional tobacco farm and tobacco control leaders were now at least aware that some engagement between the two groups had occurred.

In the meantime, the Robert Wood Johnson Foundation SmokeLess States program had issued another request for proposals, this time for projects up to four years rather than two. Virginia's tobacco control leadership, led by Rebecca Reeve, sought and received funding that would allow expansion of the dialogue between tobacco control and tobacco farm interests to other tobacco producing states through what was called the 'Southern Tobacco Communities Project' (STCP). The stated purposes of the STCP included:

- understanding and documenting how cultural, political and economic factors affect tobacco growing communities;
- identifying and evaluating new economic opportunities and strategies in tobacco communities;
- establishing relationships and seeking creative ideas and advocates for federal, state and local policy change;
- informing and enhancing the public debate on tobacco and economic issues.

Those purposes were encapsulated publicly through invitations to participate by articulating four relational goals. These goals, as articulated in STCP promotional literature, declared that participants were working to create constructive relationships among tobacco producers, health advocates, and others concerned with changes facing these families and communities, to enable them to:

- Replace inflammatory rhetoric, stereotyping, and automatic enmity in favor of civil, problem solving dialogue;
- Understand each others' needs, values and concerns;
- Identify areas of common ground and even interdependence, while acknowledging areas of difference; and,
- Work together to create realistic, sensible and sustainable options for communities and families facing pressures associated with transition.

These may have been noble goals; however, the decisive factor in tobacco farmer participation in the Southern Tobacco Communities Project was not these project goals or the limited success of the Virginia Tobacco Communities Project. Rather, it was that many farmers realized that permanent changes were occurring within the South's tobacco growing communities. Tobacco was still in decline. Both income from tobacco sales and the numbers of growers had declined significantly in recent years. Even in growers' best-case scenario from their own perspective, where American tobacco continued to be in demand as the premium tobacco in the world, many fewer growers would be working than was then the case. This change could be devastating to families and communities, particularly in rural areas where poverty and unemployment is already high.

The structure of the Southern Tobacco Communities Project (STCP)

Thus the stage was set for enhanced, meaningful dialogue. The STCP grant proposal had called for formation of a fixed-membership roundtable of farmers, researchers, community development officials, and public health advocates. Smaller dialogue groups would be formed in each of the six main tobacco-producing states, and groups would be formed according to topics as well (see Fig. 10.1).

The plan for the STCP declared four components to achieve its goals:

- A regional "roundtable" of tobacco, health and economic and community development interests to develop policy options;
- Support for initiatives in each of the six states to examine these issues on a statewide level;
- Development of a model tobacco adjustment matrix of economic development opportunities and options;
- Development of lasting, realistic, productive relationships among the differing interests to carry on the work beyond these structured activities. In addition, a Planning Group made up of leaders representing tobacco farm and public health advocates would encourage a core membership to assume leadership responsibility for the discussions and provide legitimacy for discussions and planning. That Planning Group could also assist facilitators in identifying issues of concern and fulfill an important liaison role with other members of their constituencies.

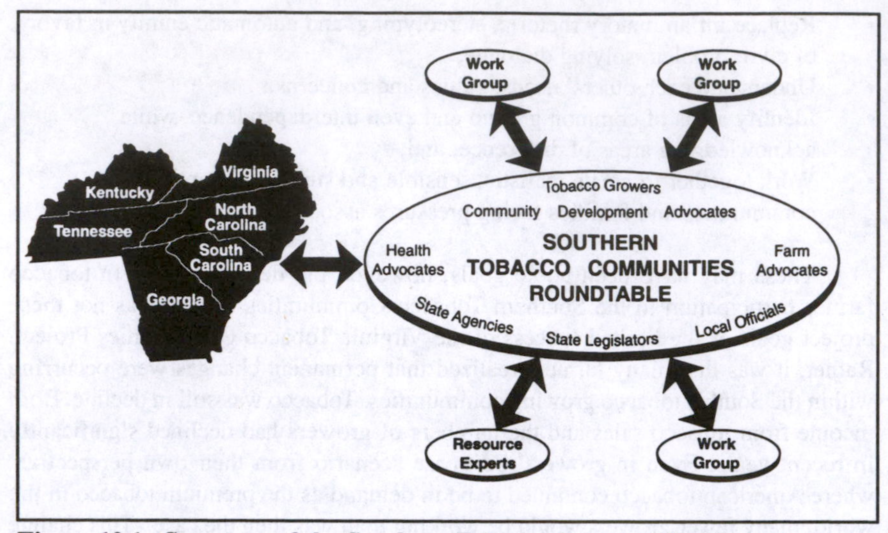

Figure 10.1 Structure of the Southern Tobacco Communities Project

As was to occur many times, however, the political sensitivity of tobacco intruded. While some people endorsed the STCP goals and were willing to meet with their presumed adversaries to explore common ground, few would commit to any more than a single meeting at a time. Fewer still had any interest in any formal roundtable membership or serving, even in an unofficial capacity, on an advisory group. And funding was only provided to support state dialogues in Kentucky, South Carolina, and Tennessee. Thus the plans needed to be modified. Accordingly, we returned to the model of the Virginia Tobacco Communities Project (VTCP), with a series of ad hoc meetings that promised to continue only if invited participants found them valuable.

The Southern Tobacco Communities Roundtables

Besides the expansion to other tobacco producing states, one more significant change differentiated the STCP from the VTCP. During the Virginia project, health interests were discussed in limited ways. The focus was on tobacco communities and their problems. For the STCP to attract continued public health participation, however, public health goals would need to be brought to the forefront.

A remarkable and unprecedented series of meetings ensued. The first session of what eventually extended to fourteen *Southern Tobacco Communities Roundtables* was held in Raleigh, North Carolina in 1997. Some 60 farmers, extension leaders, public health advocates, and community development advocates heard presentations concerning the federal tobacco program, youth access to tobacco products, and economic diversification.

Discussion varied between presentations of facts and arguments about policy. Health advocates learned that the gross income per acre from flue-cured tobacco was $3,803 for the most recent three-year period, with a production cost of $2,591. The operating income per acre for flue-cured tobacco, at $1,212, was significantly higher than corn, which had an operating income per acre of $27, soybeans at $1 per acre, cotton at $72, peanuts at $214, and wheat which lost $5 per acre. They learned that wheat and soybeans are rotation crops for tobacco and share some of the same equipment, so may actually be a little more profitable than indicated. And rotational crops are often grown simply to keep labor available and busy throughout the growing season.

They also learned that agricultural researchers studying the effects of eliminating the tobacco program concluded that it would lead to:

- A reduction in the numbers of farms and farmers, an increase in the size of the remaining tobacco farms, and a destruction of the investment in return-to-quota;
- An increase in tobacco production, at least in the short term;
- A shift in income to purchasers and chemical and fertilizer dealers because of the larger-scale production methods, with farmers working harder to have the same income; and,
- A reduction in the US and world price of tobacco by at least 40 to 45 cents per pound, with potential dramatic results on developing countries that grow tobacco.

They also learned that the typical burley farmer in Kentucky grows about two acres, which means that most farms are very small. And they heard that growers feel that they have to keep looking over their shoulder to see who's shooting them in the back. Tobacco growing is part of a heritage passed down by the family through the generations, and many growers bitterly resent having their heritage attacked.

Much of this was new to public health advocates. And almost all of what the farmers learned was news to them. They saw that at least some of the public health advocates spoke of the need to support rural communities. They learned of the difficulties in preventing youth from using tobacco products. And in what was the most significant presentation of all, they heard from one public health advocate who presented reasons why, from a public health perspective, it could be important to support the tobacco program. These included:

- Without the program, the price of unprocessed tobacco on the warehouse floor would decline. Based on studies of how cigarette price affects sales, this drop in cigarette retail price would undoubtedly lead to an increase in the use of tobacco products;
- Because of the decline in prices, resources would be redistributed from the growers to the manufacturers;

- Without the program, many small and medium-sized farmers would be put out of farming, bringing increasing economic hardship and other public health problems;
- Removing the program may negatively impact farmer capacity to invest in supplemental enterprises;
- The tobacco program actually brings together unusual partners, such as the Roundtable today. The absence of the program would reduce opportunities for dialogue between growers and health groups.

Growers and health advocates have some potential areas for agreement and commonality in their (1) concern for the physical and economic health of families and communities, and (2) risks taken in coming to the "table" to talk to each other. Growers and public health groups need to be able to disagree; they need to be able to put issues on the table, move beyond their points of disagreement, and identify issues on which they can work together.

In a most prescient statement, some participants suggested that the two groups may need each other if they are to make a political impact. Unless people are willing to work together, these people declared, the situation will not change, because no government agency will bring about change for them. To work together, people need to make a commitment to understanding, to be innovative, and to avoid stereotyping.

Results from the Roundtables: a new alliance

The first Southern Tobacco Communities Roundtable meeting was followed by a series of actions:

- The continuing series of regional roundtables, where participants could get up-to-date information from knowledgeable researchers and agency personnel and explore common ground even as the world of tobacco and tobacco policy shifted every few months;
- State dialogues in Kentucky, South Carolina, Tennessee, and Georgia, where participants could work on issues of local statewide importance to farmers and health advocates;
- Farm tours, both regionally and by state, where health advocates got to meet a variety of growers and hear of their lives, their hopes, and their worries;
- Smaller, private, ad hoc meetings where current policy matters could be discussed with assurances of confidentiality;
- Joint visits to legislators and conferences to explain the growing common ground.

Core Principles Statement Between the Public Health Community and the Tobacco Producer Community

In early 1998, the first substantial policy statement to emerge from the new alliance was released. The *Core Principles Statement Between the Public Health Community and the Tobacco Producer Community*, as it came to be called, represented a bold step and substantial risk, particularly to the tobacco producers, who were on record now for perhaps the first time as endorsing public health positions that the tobacco manufacturers – their customers – strongly opposed.

These "Core Principles" did not come easily. At times, differences around their development threatened to derail the entire Southern Tobacco Communities Project, particularly when the tobacco producer community split over language concerning the tobacco control program. However, once the Core Principles were established they gained support from nearly all the major public health and tobacco farm organizations. They provided an antidote to tobacco company propaganda and suspicion from public health advocates outside of the tobacco producing states who questioned the legitimacy of this emerging alliance.

The Core Principles asserted the need and mutual support for these important policies:

- A privately funded but federally managed *tobacco production control program,* which provides *stability* for growers, *limits supply* and sets a *minimum purchase price,* without taking public tax dollars;
- *Compensation* for quota holders, lessees, and tenant farmers as their income declines if and when tobacco use is reduced;
- Funding for *locally-controlled investment* in tobacco growing communities to diversify their economies, where local control ensures that farmers themselves have a say in how any such funding is spent;
- Effective *laws* and *enforcement* to restrict *youth access* to tobacco products, while ensuring informed adult choice;
- *Cooperation* between tobacco producers and public health advocates to improve the safety and health of tobacco production and avoid additional regulatory burdens on farmers;
- FDA authority over tobacco *products* only (off-farm) to protect public health and ensure consumer access to industry information about such products;
- Assurances that imported tobacco, which may have additives, pesticides, reconstituted leaf, and may be associated with other unregulated practices, meets the same health standards as domestic tobacco, which could improve American producers' competitive position while helping ensure a less harmful product;
- Assurances that *if* an excise tax increase is imposed on tobacco products, portions would be used for tobacco community assistance as well as public health initiatives.

Virginia allocation of settlement funding for tobacco communities and public health

Public health advocates demonstrated their support of the program in many ways since the Core Principles were signed. They provided farmers with access to policymakers not considered friendly to tobacco interests. They also successfully advocated maintenance of the crop insurance program when Congress looked as though it might eliminate the program.

In 1998, the states' attorneys-general announced a settlement with the tobacco manufacturers in compensation for Medicare expenses related to tobacco use. In Virginia, the anticipated amount was to be about $4.6 billion over 25 years. Late that year, a small group of health advocates, farmers, and legislators met to determine whether they might jointly seek portions of that funding for their needs. The public health and agricultural communities agreed to seek 50 per cent of the state settlement money for farmers and their communities and 10 per cent of the settlement money to be directed towards youth prevention and cessation. That left 40 per cent to be used for other purposes, including more public health initiatives.

Despite efforts to derail the legislation by the governor, who was closely aligned with the tobacco manufacturers, the legislation passed with unanimous votes in the House and Senate. Both sides agreed that it could not have been passed without the active support of each community. While Virginia's allocation of the settlement funding is the most striking example of the utility and power of the new alliance between public health and tobacco farm interests, other states have seen benefits as well. In Kentucky, farmers have lent their voices to media campaigns designed to reduce youth smoking, and the Burley Co-operative voted in October 1999 to endorse spending 15% of the state settlement money towards youth prevention and cessation efforts. Other states have seen support in the form of joint lobbying on behalf of tobacco and health.

A Presidential Commission

The culmination of many STCP participants' hopes occurred late in 2000 with the appointment of a Commission by President Bill Clinton, the ***President's Commission on Improving Economic Opportunity in Communities Dependent on Tobacco Production While Protecting Public Health.*** That Commission, made up of prominent tobacco farm, public health, and community development leaders, was initiated largely through the efforts of active STCP participants and included members who have been active in Southern Tobacco Communities Project activities and others whose organizations have participated regularly as well. Two Southern Tobacco Communities Roundtable meetings were held to address the Commission's work.

The final Southern Tobacco Communities Roundtable meeting was held on May 29th, 2001, just after release of the Report of the President's Commission: *Tobacco at a Crossroad – A Call for Action*. Introduced by several members of the Commission who participated in the Roundtable, the Report was intended to create a package of actions and benefits that unites tobacco farmers and public health advocates in the common goal of protecting tobacco farmers and their communities while improving public health.

The recommendations of the Report fell into three categories: the Tobacco Equity Reduction Program (TERP), a Center for Tobacco-Dependent communities, and the Public Health Proposals. The key recommendations were:

- The adoption of a tobacco equity reduction program (TERP) to compensate quota owners and growers for the loss in value of their quota assets;
- A major restructuring of the federal tobacco control program replacing quotas with permits that could not be sold or leased, to address changing economic and public health realities;
- The establishment of a Center for Tobacco-Dependent Communities to address the economic development challenges and opportunities that tobacco producing communities face;
- Providing fair and effective regulatory authority over the manufacture, sale, distribution, labeling and marketing of tobacco products by the Federal Drug Administration (FDA);
- Ensuring adequate funding for tobacco prevention and control programs by the states as suggested by the Surgeon General of the United States in an August 2000 report, including coverage under Medicare and Medicaid programs for smoking cessation.

The 30 participants at the meeting, national and regional leaders from the tobacco farming and public health communities, included several Presidential Commission members. Participants overwhelmingly supported the Report's recommendations as a package and in their entirety. While recognizing the difficulties faced in enacting the recommendations, they believed that an alliance of tobacco farmers and public health advocates could wield considerably more influence than any single group on its own. The attending Commission members congratulated Southern Tobacco Communities Project participants for laying the groundwork that made this unified effort possible.

Conclusion: will this alliance last?

Looking back at the work of the Virginia Tobacco Communities Project and the Southern Tobacco Communities Project, many of these outcomes seem inevitable. But that is not the case, and the future certainly remains unknown. At times, their own friends and allies have criticized growers and health advocates for working together. The ongoing effort by some tobacco manufacturers to undermine the tobacco program by switching to direct contracting with individual farmers, if successful, would mean that farmers would lose the independence that the program provides. Farmers already report threats to their livelihood from leaf dealers and manufacturing interests concerned with this alliance, and a move to contracting could end any future farmer/public health coalition.

Another obstacle is the natural competition for resources, attention and power *within* the health and grower communities. No one entity speaks for the entirety of either community, as has been made abundantly clear by the proposed settlement. Each side, quite naturally, has internal divisions, as situations, constituencies, and organizational imperatives differ. When working in coalitions, it is important to remember that tensions likely will occur in direct proportion to the amount of work, personal commitment, and meaning everyone has invested in their work. Here we are dealing with one group's culture and livelihood, and another group's mission to promote health and save lives. Plans for activities which might influence others' projects should be shared as early as possible. People need to learn to speak directly to others of their concerns. A growing cycle of distrust, miscommunication and misunderstanding is the inevitable result otherwise.

In addition, the volatility of the political arena, as has been evidenced many times during the past decade, changed the dynamics of the discussion such that development of a deliberate, long-term agenda, and building the necessary grass-roots and leadership support for such an agenda, was impossible. At the same time it provided some substance and urgency for interaction. A willingness to allow flexibility with the *process* has been necessary to achieve desired *outcomes*.

And, as might be anticipated, these discussions do not always proceed smoothly. After all, we are dealing with one group's culture and livelihood, and another group's mission to promote health and save lives. Where knowledge of the other is lacking and there exists doubt about others' motives, it takes little to upset a relationship. Neither is there any established institutional vehicle that brings the two sides together on a regular basis to address potential difficulties.

Growers are under considerable pressure to align themselves absolutely with manufacturers and other tobacco interests. They recognize that they must maintain working relationships with their customers. But potential benefits to all sides from continuing the alliance are substantial. Farmers benefit by:

- Improved understanding of the values, goals and actions of the public health community;
- Development of political alliances with a powerful player (public health groups) in the political sector, bringing a counterbalance to their dependence upon manufacturing interests;
- Support for economic development in rural, tobacco producing communities;
- Improved public relations and better standing in the public's eyes as the occurrence and outcomes of such dialogue are revealed.

Benefits to the health community are equally substantial:

- Improved understanding of the values, goals and actions of rural, tobacco growing families and communities;

- Development of political alliances with a key player (tobacco and farming interests) in the political sector;
- Improved relations with a group with many decades of experience negotiating with the tobacco manufacturing industry;
- New opportunities to improve public health in rural communities;
- Improving understanding of tobacco markets, overseas operations, and strategies to negotiate with tobacco interests;
- Opportunities to get important support for public health initiatives.

As to the broader lessons that can be drawn, forums of this sort for productive dialogue between adversaries, or perceived adversaries, who have not spoken with one another before, have many functions and benefits.

They serve to break the ice, to develop a capacity for talk where quite literally none existed before. It sounds simple, but where no forums exist for such talk it is virtually impossible for such talk to occur. Once the ice is broken, it stays broken.

They allow for new learning. For tobacco growers and the public health community there has been an incredible, albeit understandable, amount of dependence on ignorance or stereotype about each others' beliefs, values and goals. With some exceptions, the health community has been generally ignorant of the basis of tobacco production and the economics of rural tobacco producing communities. Tobacco growers have generally been strongly influenced by manufacturing companies and media distortions of health advocates' intentions and behavior. Direct, face-to-face discussions invariably result in new, more accurate understandings.

They weaken the extreme and empower the reasonable. When all communication is conducted through filters of media and third-party reports, what is most likely to come through is what is shouted the loudest, what excites the greatest reaction, and what is most inflammatory and "newsworthy." Facilitated discussion in particular, when done properly, focuses on providing all interests an opportunity to speak without interruption, to be heard with respect, and to listen with attention. By showcasing voices and ideas which have not had currency, and by modeling processes of respectful, candid and productive dialogue, the extremes tend to stand out as extreme and the reasonable as worthy of consideration.

They create empathy and therefore recognition of common values, goals and concerns. For example, those growers and health advocates who have been most actively engaged in discussions with one another have realized that they assume a considerable risk of ostracism and criticism within their own constituencies for meeting and working together.

They allow for productive action even in circumstances where some differences continue. Because parties realize that they can talk to each other with civility, they can work together when appropriate on certain issues while opposing each other on other issues.

Areas of joint interest continue to be discovered as discussion proceeds. If trust can be developed that parties will not abuse the demonstrations of candor and honesty, dialogue will continue to be productive.

At the personal level, people who used to be considered opponents learn that they do share many values, such as hard work, community involvement, personal and social responsibility, and a concern for youth and community economic, physical and mental health. Both groups benefit at the personal level by recognizing that people who used to be considered opponents in fact share many values, such as hard work, community involvement, personal and social responsibility, and a concern for youth and community health.

As the STCP wound down in late 2001, STCP participants contributed significantly to a new group that has formed to carry out the work of the President's Commission, the *Alliance for Health, Economic and Agriculture Development* (AHEAD). That organization has taken on some of the tasks envisioned for the Southern Tobacco Communities Partnership, with a focus on policy advocacy. Continued benefits will depend upon good faith efforts on all sides to reach out to others in ways that bring forth new relationships and new ideas.

Notes

1 Six states produce the bulk of American tobacco, although it is grown in a dozen other states in much lesser quantity. These six are Kentucky, North Carolina, South Carolina, Virginia, Tennessee, and Georgia.

2 Continuation of the tobacco program of price supports and quotas has been the highest priority of most tobacco growers. The tobacco program was created to give back to the growers some control over their destiny, and to create stability for the industry.

The *tobacco production quota* is an important part of the federal tobacco program administered by the U.S. Department of Agriculture (USDA). The program includes a system of quotas and *price supports* by which the production and sale of U.S. grown tobacco is conducted. The tobacco program was intended mostly to serve two functions:

1) To give back to the growers some control over their destiny by removing their absolute dependence upon leaf purchasers and manufacturers;

2) To create a certain level of price and production stability for all elements of the industry.

In the U.S., tobacco farmers must either own or lease the right to produce and sell tobacco. Due to fluctuations in the market, the amount a farmer may grow may vary from year to year. In addition, a minimum purchase price is set. The price offered by leaf purchasers must exceed the minimum price set by USDA in order to be sold. For tobacco brought to the warehouse for auction, if nobody bids the minimum amount, the tobacco will be purchased by one of nine co-operatives responsible for that particular type of tobacco at the minimum purchase price and the farmer will return home with a check for that amount in hand. The co-operative is lent money by the Commodity Credit Corporation to purchase that tobacco, and later repays the federal treasury with interest with fees collected from farmers and leaf purchasers, called the *tobacco marketing assessment*. That unsold tobacco is eventually marketed (usually at a loss) by the co-op. That quota amount is set each year for each particular type of tobacco. That amount is determined by a variety of factors, including the purchase intentions submitted before each year's growing season by the manufacturers, the amount of tobacco left unsold and taken under loan by the co-ops from previous years, and the three-year average of exports. Farmers who produce more tobacco than their allotment may not bring it to market until a following year.

Many farmers and others believe that the amount of manufacturers' purchases is determined not only by their production needs but by the need to exert control over tobacco farmers' political choices.

3 Confidential personal communication.

4 Personal communication.

Chapter 11

Rural Action: Participatory Planning for Healthy Communities in Appalachian Ohio

Christopher S. Rice and Carol Kuhre

Introduction

Rural Action is a non-profit, membership-based Non-Governmental Organization which promotes social, economic, and environmental justice in Appalachian Ohio. Headquartered in Athens County, Ohio, it currently has projects in 14 counties and makes an impact in most of Ohio's 29 Appalachian counties. Rural Action's vision, which, as we shall later see, grew out of the participatory work of citizens who helped found the organization, is to develop model strategies that involve a broad base of citizens in building environmental, economic, and social sustainability in the region. Contrary to earlier approaches to rural development in Appalachia, such as the War on Poverty of the 1960s, which were focused on the political, social, and economic *deficiencies* of the region, Rural Action's work in Southeast Ohio is *asset*-based. Following the principles of community development advocated by planners such as Kretzmann and McKnight (1993) and Shuman (1998), Rural Action's principles focus on the region's assets to: (1) promote economic and social development; (2) restore and protect the environment for future generations, and; (3) revitalize each community in its own unique way.

"It's all a question of story," said philosopher/theologian Thomas Berry. "We are in trouble now because we do not have a good story." Rural Action is actively reshaping the old story – a set of beliefs that has formed current Western culture: the possibility of unlimited growth; the Enlightenment belief in the inevitability of "progress" (which is always seen as positive); and the fetishization of technology, believing that it will always be able to solve our toughest problems. In its place, Rural Action offers a new story – the Strategy for Rural Renewal, a comprehensive approach to rural community development. The Strategy for Rural Renewal provides the foundation for Rural Action's integrated development strategy, a participatory approach which builds on the assets of the region and individual communities to develop social capacity and civic democracy. The result is the realization that long term solutions for Appalachian Ohio must be inclusive of a broad spectrum of citizens to both restore and preserve environmental resources and revitalize the region's economy. Like all good stories, however, Rural Action's story is a result of the events of an older, longer story, and is intimately bound up in its telling. To understand the story of Rural Action, one should first know a bit about the story of political resistance and community development in Appalachia.

Appalachia has long been a site of cultural, economic and political contestation. As Fisher (1993, p4) has noted, recent Appalachian scholarship has recognized the importance of the region's history of resistance and struggle, realizing that it "has most frequently occurred in struggles to preserve traditional values and ways of life against the forces of modernization," which, we would argue, in its late-capitalist form can be seen as the phenomenon of globalization. Fisher also notes that, concurrent with the colonialist analysis of Appalachia in the late 1960s and early 1970s, an "explosion" of community organizing centered around "hundreds of citizens groups" emerged, though largely constructed around the struggle against a single issue, such as strip-mining, or more recently, acid mine drainage and mountaintop removal. However, the attrition rate for these grassroots organizations in Appalachia has been particularly high. As Fisher (1993) observes,

> Most [single-issue, grassroots organizations] have proved unable to establish continuity or to see beyond the immediate crisis. These single-issue groups have worked together from time-to-time, have helped create local leadership, and have won important victories; but most have been short-lived, disappearing quickly once their issue was resolved. (p8)

Echoing Fisher's themes, Mary Beth Bingman, herself a participant in this mode of resistance, asserts that while single-issue organizations and their efforts may win important, individual victories, this mode of political resistance cannot of itself bring about fundamental change (Fisher, 1993, p6).

More recently, successful change in the region has been brought about by the "establishment ... of thriving and influential multi-issue, membership-driven organizations" (Fisher, 1993, p8), a group in which we would firmly place Rural Action. These dynamic organizations are essential for the creation of what Evans and Boyte (1992) have referred to as "free social spaces," grounded in everyday life, in which "people can learn democratic values and leadership skills, obtain alternative sources of information about the world, form a coherent pattern of group identity and a vision of the common good, and act on their values and beliefs" (Fisher, 1993, p319). In short, these free spaces allow for the construction of civil society and the development of social capital necessary for participatory governance. Moreover, they channel those social forces for political development and resistance by providing discursive space "where 'people's history' may be connected to a systematic critique of the political economy; where participants can begin to see the connection between their concerns and those of other exploited people;...and where people can start to envision new alternatives to the world in which they live" (Fisher, 1993, p329). It becomes our task, then, to show the history and struggles of Rural Action *qua* 'free space' as parallel to, and intertwined with, this history of resistance and the struggle for sustainability in Appalachia.

A brief history of Rural Action

Rural Action began in 1982 as the Appalachian Ohio Public Interest Campaign (AOPIC), the first rural branch of the Ohio Public Interest Campaign (OPIC)/ Citizen Action organization. The group's initial raison d'être was 'to take on issues in this poor rural area [Southeast Ohio] that other groups, particularly government agencies, could not or would not address' (Couto, 1999, p122). The group was largely reactive, responding to Reagan administration budget cuts in human services programs, as well as to local environmental issues, such as water and subsidence problems brought on by longwall and strip coal mining, among others. Carol Kuhre was on the founding committee of AOPIC, and was intensely involved with the organization's Board around 1982-1983. However, demands on her time as Co-Director of the United Campus Ministry at Ohio University, as well as international travel to Africa and the Philippines, took her away from the nascent organization until 1990. During that time, AOPIC discovered both the benefits and disadvantages of belonging to a larger, state-wide organization. Though the group certainly benefited from expert staff assistance made available through OPIC, as well as the larger groups cache with, and connections to, large funding sources, after more than seven years of good work they realized that carrying out a state-wide agenda, important as it was, left AOPIC with neither the time nor the energy to adequately address important local issues. The AOPIC board was also unhappy with the policy of OPIC to canvass the southern part of the state but to not share the revenue with the local Appalachian office. Growing increasingly weary of the costs of their affiliation outweighing the benefits they received, the local organization eventually decided in 1989 to 'leave the nest' of OPIC and go it alone (Couto, 1999, pp122-23). The group then evolved into the Appalachian Ohio Public Interest Center (AOPIC) in order to establish themselves as a center for activity in the Appalachian region and not a part of a state-wide campaign.

However, the newly-independent organization quickly found itself cast adrift in the funding world, having lost the connections afforded them by their affiliation with OPIC. Eventually, AOPIC was down to one funder, the Commission on Religion in Appalachia (CORA), which continued to provide AOPIC with a very modest operating budget. In addition to financial stresses, the group was also facing a crisis of morale among its affiliated groups. AOPIC had spun-off or affiliated with 23 smaller citizen's groups, such as Citizens Organized Against Longwalling (COAL). By 1990, several of these smaller groups had achieved their goals, and wanted to go back to living a normal life. Many others had not achieved their goals, leaving some with the desire to continue their struggles, while others were ready to "pack it in". A large number of the citizens who were participating in these small groups were tired from years of fighting and burned-out from the ever-present need for fundraising.

When Carol Kuhre returned to Athens, Ohio, from Africa in 1989, the director of AOPIC was in the midst of deciding whether to stay with the organization in light of the funding instability. The board was also reflecting upon the future of the organization. Kuhre, then employed as an organizer with the Appalachian People's Action Coalition, worked with the board to put together a two-step assessment plan that included site visits to the issue-focused groups with whom AOPIC had worked in the past eight years and a four month process whereby all groups would be invited to joint meetings to evaluate and plan for the future. Twenty-three organizations and selected individuals (53 in all) participated in the planning process that resulted in a new vision for the region and the organization.

AOPIC learned from the assessment process that the AOPIC staff and the grassroots volunteer leaders had latent visions of what they wanted the region to become but because of the intensity of their issue-based campaigns they had not found the time to make these visions manifest. The planning process allowed these citizens to begin to articulate a vision of what a sustainable Southern Ohio might look like, creating a synergy among these citizens that had been unable to emerge while they had been focused on their individual issues.

From these early brainstorming and evaluation sessions, a mandate emerged to network the groups more closely together and to form committees to work out the details of the new vision. During a transition period lasting about three years, AOPIC volunteers and part-time paid staff did intense organizing that resulted in the formation of committees of sustainability around agriculture, arts and heritage, economic development, education, energy, forests, health, housing, human needs, recycling, and pesticide reform. These committees were the incubators for current program initiatives.

From the beginning, and still true, are mandates for each committee to have five paid active members, to create a strategic plan for its work, and to record and post all meeting/conference proceedings with the central office. Rural Action staff, familiar with the organizing work of Saul Alinsky and Cesar Chavez, realized that some small financial investment in the organization was/is important to encourage a sense of being linked to the group and its future and of ownership in its work The rules also have a dual purpose of keeping the committee members engaged and responsible to the organization, as well as forcing the staff experts who work with these committees to remain engaged with their committees and keep their initiatives citizen and member-driven rather than staff-driven, promoting the maintenance of participatory decision-making.

Committees are very diverse, reflecting the "clustering" that went on during the planning process. For example, the agriculture committee includes farmers, would-be farmers, gardeners, marketing specialists, extension agents and academics interested in rural and agricultural issues. Two key principles of sustainability are those of diversity and integration. In its committees, collaborations and board structure, Rural Action attempts to bring as many diverse parties to the table as possible, always attempting to seek common ground among them in order to create social good.

Healthy communities and sustainability – The Rural Renewal Strategy

> If ecological good sense is to prevail, it can only do so through the work and the will of the people and of the local communities. (Wendell Berry, 1992)

Sustainability, as a concept, has been a fundamental part of Rural Action's work and identity, almost from the very beginning. When AOPIC became Rural Action, one of its first VISTAs (Volunteers In Service To America) was assigned to build and formalize this dense network through a series of conferences centered around themes of sustainability, such as biodiversity, bioregionalism, sustainable agriculture, etc. Sponsored by the Rural Action Network, a network of organizations in sync with the Rural Renewal Strategy with whom Rural Action places VISTA volunteers, these conferences were structured so as to be cross-sectoral and inter-organizational from their inception. In addition to the organizations that had been networking since 1990 and the citizens that Rural Action had been organizing since then, the meetings also included representatives from local and state governmental organizations such as the Ohio Department of Development, the Governor's Office on Appalachia and County Commissioners. These meetings were remarkable because of the incredible cross-pollination between governments, experts, and citizens present at the meetings. Though the initial purpose of the series was to make the members of the network more conversant with international and national literature on the subject of sustainability, something rather different resulted. Instead of accepting a formal, government-endorsed definition of sustainability, the participants began to deconstruct the concept of sustainability in order to adapt it for the local context. Some participants questioned the use of the word, for fear that local citizens would not understand it. Some asked if, given its many contradictory uses, the term "sustainability" even meant anything.

What emerged through the course of these deliberations between citizens, academics and government officials was that sustainability, for Southeast Ohio, meant robustness, or more specifically, health – healthy community, healthy economy, healthy environment, healthy individual. As Michelle Ajamian indicates, use of the term healthy 'relates to the idea of holistic health that looks at the organism as a whole and sees that fixing one part of what is wrong will never create true health' (Ajamian, 2002). Subsequently, Rural Action (as the increasingly dense network came to be known) developed criteria for sustainability in Appalachian Ohio, though sustainability language is often not used, in favor of the 'healthy' paradigm. One drawback to the process, however, was that these initial conferences on sustainability had largely involved citizens who were academics (Athens, Ohio is a university town) and government agency personnel. Few lower-income members had become involved. To rectify this, Rural Action organizers developed a second series of meetings on sustainability focused on involving lower-income members of the organization and the community. The result was a publication, *Building Healthy Communities*, which made sustainability concepts and indicators accessible more broadly to the Rural Action membership and community. Rural Action continues to grapple with the idea of sustainability, and continues to work toward demystifying the concept, as well as striving to develop educational tools to educate its membership and VISTAs on sustainability. Rural Action's Strategy for Rural Renewal works toward a sustainable future for Appalachian Ohio through three primary initiatives which utilize the concepts developed from these series of meetings.

Strategic initiative 1: The Sustainable Communities Initiative

The Sustainable Communities Initiative posits that sustainable community development requires active citizen participation. Rural Action's community development work builds the capacity of communities to understand their past, map their present situation, vision their future, and develop strategies for rural renewal. In view of the history of political resistance and struggle in Appalachia, it should not be surprising that the concept of 'community' would be used as a focus for sustainability by Rural Action. Democracy theorist Benjamin Barber claims that 'when we use an inescapably political term like civil society, its ideal normative meaning as given by certain democratic and civic ideals is inextricably bound up with various civic attitudes and practices that surround it in our lives' (Barber, 1998, p13). Fisher (1993), observes that most grassroots NGOs in Appalachia have been 'community-based'. This is an important part of the efforts to establish a stronger civil society in the region, to place the notion of 'community' firmly at the center of the locus of participatory governance, for 'community', as Zygmunt Bauman states, 'stands for the kind of world which is not, regrettably, available to us – but which we would dearly wish to inhabit and which we hope to repossess' (Bauman, 2001, p3). Thus, in the efforts to establish a strong civil society and spaces for democratic, participatory decision-making, community, like the relationship between health and sustainability, has come to be an appropriate and meaningful term for this in the context of Appalachian Ohio. Community, if we read Baumann correctly, becomes at once the desired end and preferred modality of achieving sustainability arrived at in a participatory manner.

The development of a new generation of leaders for Southeast Ohio, focusing on citizens who have not been given the opportunity for leadership in the past, is at the heart of Rural Action's community development work. Rural Action recognizes that the expansion of participatory decision-making may cause existing leadership to feel threatened, and has organized the program to assuage the fears of existing leaders that Rural Action's leadership development work will undermine their authority. More importantly, because Rural Action uses an asset-based approach, existing leadership is seen as part of the local asset base, not as an obstacle. Existing leaders are invited early on to become part of the process by mentoring emerging leaders, allowing them to feel that Rural Action is working with them to enhance the total pool of leadership available to the community. Seeing the benefits to such an approach, government agencies at a national and local level, such as the Appalachian Regional Commission, the Governor's Office of Appalachia, the Buckeye Hills-Hocking Valley Development District and the Ohio Valley Regional Development Commission, have partnered with Rural Action in this successful program. In addition to partnerships with government agencies, Rural Action is also supported by the Catholic Campaign for Human Development, which funds much of Rural Action's community organizing work.

The asset-based Partners in Leadership Development Program (PLD) and the Community Organizing and Support Initiative (COSI) consist of three phases. In the Pre-Development/ Groundwork phase, Rural Action first hires citizens from local communities that have been identified as potential sites for action to work as VISTA field organizers. These organizers had to have grown up in the area, have family there, or possess some other intimate connection to the community. Although there have been a few mistakes with this process, such as the hiring of organizers who seemed to meet these criteria but unbeknownst to Rural Action were not, for one reason or another, accepted in their communities, this use of local citizens as VISTA field organizers has been a generally successful and important part of the overall PLD/COSI approach.

In this first phase of the PLD program, the VISTA field organizers spend considerable time and effort, usually from 8-12 months, in the selected community developing a solid relationship of mutual trust with citizens, local officials, and government agencies in the area working on community development issues. PLD Coordinator Candi Withem states the importance of this phase of the program: "By familiarizing themselves with the history, culture, and human, social and physical assets of the community, these organizers build the trust and support of local people" (Rural Action, 1999, p1). This period of familiarization allows the organizer to identify potential participants for the program. Importantly, the organizer develops a series of inclusive activities to obtain the participation of a broad cross-section of community members. Candi believes that "some of the best efforts of our organizers come from just talking and working along with people; sharing their memories, listening to their dreams, envisioning a better future for their families" (Rural Action, 1999, p1). Eventually, some of these citizens and previously-identified community leaders are recruited to form "Action Teams" to pursue local community development goals. Since PLD/COSI began in 1995, Rural Action has developed Action Teams in 18 small communities.

The PLD's second phase, the Workshop Series, provides leadership and community development skills training to Action Team participants through monthly workshops. Participants acquire leadership skills, such as conflict management, and meeting facilitation, as well as asset-based community development skills, including asset-mapping, visioning, strategic planning and grant writing. Because Rural Action recognizes that each community, each Action Team is unique, with its own sets of gifts and ideas, the PLD training process retains a high level of flexibility, allowing VISTA organizers to modify the workshop schedule to suit the needs and pace of each team (Rural Action, 1999, p2). Over the course of the PLD process, each Action Team develops a strategic plan, with specific action steps, for a community development project strictly of their own choosing. Following the workshop series, each Action Team receives a micro-grant, or 'seed money', of US$400-600 to put their designated project into action. During this phase, Action Teams are also strongly encouraged to continue their community development work by networking and collaborating with other local NGOs and government to increase their capacities and leverage additional funds. This phase of the program concludes with a celebration in which the Action Teams present the process and results of their projects and their experiences of the PLD program. All Action Teams and PLD staff from Rural Action, as well as local government officials, regional community development professionals, community leaders, other local citizens and the media are invited to attend. This celebration, serves as a powerful builder of local and regional social capital, and as the beginnings of a 'community of communities' (Daly and Cobb, 1994; Held, 1995).

The Follow-Up Activities, in which the COSI emphasis shifts to support and assistance to the fledgling Action Teams, are the third and final phase of the PLD program. As Candi Withem puts it, 'at this point we step back and encourage the Action Teams to spread their wings and fly' (Rural Action, 1999, p2)' Field Organizers continue active work with the Teams, but in more of a supportive role. Rural Action also continues to provide technical support, such as networking opportunities, computer troubleshooting, grant writing assistance, etc., with the assumption that the training the Teams and their members received in phases one and two of the PLD program has developed their capacities to set their own directions and maintain their own level of participatory decision-making and community development work. Rural Action is currently assessing the 18 Action Teams involved in the Community Organizing and Support Initiative (COSI) in an effort to understand where each Team is in terms of its development, where they are seeking to go next, and how Rural Action might help them to further these goals.

A second important aspect of the Sustainable Communities Initiative has been the Arts and Cultural Heritage Program. Through traditional music preservation, storytelling, the collection of oral histories, the creation of community murals, poetry, drama activities and community festivals, Rural Action uses the arts and cultural heritage of Appalachian Ohio as a tool for community organizing and revitalization. The Arts and Cultural Heritage Program projects, which the organization refers to as 'art that works', help Rural Action to build trusting relationships with local citizens and leaders, cross traditional boundaries, encourage new, non-traditional leaders, and strengthen community and organizational networks. Ultimately, these efforts are meant to aid in the encouragement of participatory decision-making and the stimulation of creative sustainable community development.

This approach stems from Carol Kuhre's experience with community development and mobilization during her stays in Africa and the Philippines during the 1980s. There, she observed that, in many ways, the arts, such as storytelling, were employed as one of the first tools of community organizers, making arts and culture the 'front line' of defense (or offense) in community development. When Kuhre returned to Southeast Ohio and became Director of Rural Action, she was determined to incorporate the arts, as much as possible, into every program area of the organization. The philosophy behind this holds that culture and the arts serve to bring people together and not only allows them to have a cathartic way of dealing with the pain of the old stories of domination and loss, but it also allows them to step outside themselves and develop a hopeful vision of the future. For Kuhre, this is a very *healthy* way for communities to operate. Thus, the collection of oral histories, and the interpretation of those histories into murals, music, and other forms, becomes a healthy way for citizens to come together, developing the social capital and political organization to take on more substantive issues such as housing, sewers, or economic development. This particular way of coming together encourages citizen participation in decision-making that affects their daily lives, rather than leaving them to 'experts' who, while they may know all about the management of sewers, forests, etc., may not approach the issue in a way which is meaningful and 'true' to the local community.

The best example of Rural Action's use of local culture and the arts in participatory decision-making has been the Community Murals program. Rural Action has worked with small communities in the region to develop eight very different murals. These murals are the representations of local histories, culture, landmarks, and the hopes of their citizens for the future. The process begins with the work of a field organizer who engages a small community in a process of collective recollection of the community's past – its key events, landmarks, architecture, and citizens. This is frequently done through a series of 'potluck' dinners and meetings where citizens are given the space to collectively remember what their community has been and envision what it might become. Importantly, local youth are also encouraged to participate with the elderly of the community in this process, facilitating the transmission of important cultural capital, such as oral history, between generations. Local youth have also proven instrumental in the actual painting and maintenance of these murals.

In the Trimble Mural, painted on the wall of the cafeteria of the local high school, one can see the cultural memories of a terrible flood, of an elderly quilter whose products were instrumental in early community fundraising, of a broken town clock which resisted all the best efforts of 'expert' clockmakers to fix, yet was ultimately repaired by a local tinkerer using a spare part from a worn-out washing machine. It is the recalling of a history of hardships and the power of the citizens of the community to overcome them through their own methods which creates the social capital that Putnam (1993) and Couto (1999) indicate to be necessary not only for community economic development, but also for meaningful and lasting participatory governance.

Understanding that Rural Renewal is closely tied to public school education, Rural Action has recently extended its Sustainable Communities Initiative to include a Rural School and Community Organizing (RSCO) Project. This project comes out of Rural Action's belief that schools are centers of community, and building healthy schools strengthens every part of the community structure. As with Rural Action's other organizing work, RSCO uses an asset-based approach, recognizing the inherent skills and abilities of local people in building a strong future for themselves, their children, and their community. RSCO's mission is to organize citizens to develop strategies for strengthening rural communities through their schools. RSCO has four goals:

- to organize groups of citizens in school districts in Southern Ohio who care about schools to become more knowledgeable about education issues and develop the skills to influence decisions at the state and local levels;
- to develop a broad understanding of school funding and school facilities issues and policies in Ohio and a particular understanding of how these affect schools and communities in Southern Ohio;
- to develop and advocate for education related policy reforms that will benefit rural schools and their communities, working with individual citizens and organizations concerned about education in Ohio;
- to develop and disseminate training and outreach materials.

RSCO believes in developing the assets inherent in every Appalachian Ohio community – including schools – and that community members, in concert with their schools, should guide the process of school and community renewal. Believing that real, effective problem-solving comes from people who live in a community, RSCO strives to work in partnership with all members of a community – youth and adult, professionals, government officials, school administrators, the wealthy, the poor, and the in-between – to build coalitions to move the larger process of revitalizing Appalachian Ohio. Recognizing that this vision can only be achieved if all of these parties remain involved and connected with one another, RSCO is working to build long-term partnerships centered around local schools.

Rural Action is also involved with asset-based youth programming. This process begins by identifying the values and assets youth need to succeed and working with communities to develop programs that make them available to young people. Rural Action's latest project in this area is Youth Act, which focuses on a youth grantmaking council that will design a grantmaking program which combines community service with grantmaking and leadership development. This youth-created, youth-driven program uses service learning as a tool to teach philanthropy to the young people of the region, with adults serving as mentors, guides, and allies as the youth learn by doing and contribute to their communities in a safe, supportive environment. They will gain essential leadership skills, knowledge of concepts and skills in the field of philanthropy, as well as marketing skills as they learn to approach potential funders and 'sell' their project. Importantly, the funded projects will give something back to the communities, giving local youth the satisfaction of making a positive and tangible impact on the places where they live.

Strategic initiative 2: The Sustainable Economies Initiative

Conventional economic development in rural areas of the United States has relied heavily on recruiting outside industries to locate in a community, the exploitation of natural resources, and other means of bringing outside dollars into the local economy. Sustainable community development, however, requires a new way of looking at the creation of wealth. Rural Action has developed an asset-based approach to sustainable economic development which seeks to keep more local dollars local, use resources sustainably, and increase local ownership and expand local ownership options (*Rural Action*, 2001). The committees of this initiative have allowed citizens to gain participatory power over the governance of important local issues like agriculture, forestry, and waste management and recycling.

The Sustainable Economies Initiative's earliest program was ReUse Industries, which resulted from the efforts of the citizen members of the Recycling Committee. One area of focus during the four-month series of evaluation meetings was on discovering regional assets. Discussions in these meetings repeatedly turned to the large amount of trash in the region, and the possibilities for turning this into an economic asset rather than viewing it as an environmental deficit. Carol Kuhre soon contacted a national-level NGO, the Institute for Local Self-Reliance, which had expertise and experience in 'turning trash into treasure'. With this assistance the Recycling Committee was able to formulate an action plan to turn recycling into an economic opportunity in the region. After receiving a planning grant from the Ohio Environmental Education Fund, Rural Action went out into the five surrounding counties and networked with county commissioners, trustees, and economic development committees of the chambers of commerce to enlist them in the effort. These officials, academics and citizens held a series of meetings over the next two years to talk about opportunities for economic development through recycling in Appalachian Ohio. These meetings provided a catalyst for several projects in the region, the most successful of which was ReUse Industries, a community-owned non-profit corporation with two subsidiary businesses. This project was such a success that it was 'spun-off' from Rural Action as a separate organization in 1996. ReUse Industries remains perhaps one of the best examples of the cross-sectoral, inter-organizational participatory decision-making work at the core of Rural Action's approach.

The Good Food Direct project was developed by Rural Action's Sustainable Agriculture Committee to link the region's food consumers and producers through a subscription catalog. This method, also known as Community Supported Agriculture (CSA) (Shuman, 1998; Imhoff, 1996; Kittredge, 1996), allows local farmers to minimize financial risk by receiving investment from subscribing consumers up front, while educating consumers about the connections between nature and flavor, regional weather patterns and food availability, as well as the value of community and high-quality, nutritious food. Direct marketing efforts of CSA programs like Good Food Direct enable very small growers to reach large numbers of customers and keeps community dollars circulating locally, helping to support a sustainable and regional food economy (*Rural Action*, 2000). Moreover, Good Food Direct program has enabled a form of participatory governance in an area previously unavailable to local citizens: the local food economy. The new economic democracy resulting from this program has given some local farmers a voice in the manner of their products' distribution, allowing for more favorable economic returns, and enabling them to maintain their chosen lives as family farmers. Consumers also are empowered. By being offered a real choice as to the quality of their produce and the methods and location of its production, consumers may act as citizens to build the type of economy which strengthens their community rather than supporting transnational agribusiness concerns. Through Good Food Direct and the Sustainable Agriculture Committee Rural Action has collaborated with other local NGO's, such as the Appalachian Center for Economic Networks (ACEnet) and the Community Food Initiative, allowing Rural Action to expand its network of citizens and farmers, as well as the range of programs and opportunities for citizen participation in the region.

The Sustainable Forest Economies Project was developed by Rural Action's Sustainable Forestry Committee to offer workshops and technical assistance on sustainable timber management and the cultivation of high-value Non-Timber Forest Products (NTFP's), such as mushrooms and medicinal herbs. The Sustainable Forestry Committee selected NTFP's as the major focus of the Sustainable Forest Economies Project in line with Rural Action's asset-based approach. Medicinal herbs such as ginseng, cohosh and goldenseal are significant local assets. The current over-harvesting of these herbs was seen by the committee to be not only a threat, but also an opportunity to show the need for restorative forestry in Appalachian Ohio, much as the Recycling Committee saw the trash problem in the region as an opportunity for environmental renewal and economic development. By networking farmers, stakeholders and citizens possessing an alternative framework for economically beneficial, yet ecologically sustainable, forest management with government forestry agencies and university agriculture extension agents, Rural Action strengthened the potential of local citizens to positively impact forest management policy through participatory decision-making. Much of the Sustainable Forest Economies Project work has involved the organization and networking of herb growers and educating them as to the possibilities for NTFP-based economies in the region. VISTA organizers have gone into several communities to hold workshops with growers, resulting in the formation of the Roots of Appalachia Growers Association, an Action Team-style organization for NTFP work. Rural Action has also coordinated a Forest Congress which brought together community members from the timber industry, environmental organizations, governmental natural resource and recreational agencies, landowners, academics and researchers to discuss and find common ground on issues such as forest management, sustainability, eco-system management, development, and more. Further involving citizens with the management of local forest systems and their potential as engines of sustainable development, Rural Action, with the assistance of the Loka Institute, developed a Citizen Science Council (CSC) to do citizen research and experimentation on woodland herbs in order to share information about growing techniques and to encourage experimentation by the growers.

In addition to supporting local knowledge for sustainable forestry, Rural Action is beginning to incorporate the arts into its NTFP-led sustainable economies work as an organizational tool. Rural Action continues to explore the collection of oral histories from local growers concerning the historical uses of herbs in the region. Rural Action also wants to explore the potential for craft activity using resources from the forest understory. The potential for forest-based sustainable economies seems vast indeed. Although the Sustainable Economies Initiative will soon have an arts component to it, Rural Action is not an 'arts organization'. However, Rural Action believes that the arts can be used to enhance, enliven, brighten, and show citizens the way in all of the organization's program areas, including the Sustainable Environments Initiative.

Strategic initiative 3: The Sustainable Environments Initiative

The boom and bust cycle of an economy built on extractive industries such as coal mining and logging has left a legacy of environmental and economic impoverishment throughout Appalachia. Many streams in Appalachian Ohio have been damaged by acid mine drainage from abandoned deep mines and mining waste left by mining operations from the early 20th century. The extreme acidity of these streams, in some places the pH reaches 3.0 to 3.5 (Bernard and Young, 1997, p170), has left parts of them devoid of life and has greatly reduced the natural beauty of the area. Because of this problem, Rural Action has chosen to pursue environmental restoration in the region along watershed lines, guided by three principles:

- watershed residents must be involved in both the planning and clean-up activities so that they will feel ownership of the newly clean streams;
- collaboration and partnership with federal, state and community agencies is vital to both the process of restoration and the outcome; and,
- long-term solutions to environmental degradation must be found and are to be favored over 'quick fixes' (*Rural Action*, 2001).

Since 1994, Rural Action has been involved with several watershed restoration efforts in the region, including the formation of several watershed groups and the support of others. The process began, like other Rural Action projects, with a local need, citizen input, and a motivated VISTA. The result is the Monday Creek Restoration Project.

The Monday Creek Restoration Project (MCRP) is a partnership of over 20 organizations and government agencies working with local citizens to restore the Monday Creek watershed. It has played a pivotal role in challenging governmental culture and assumptions about the possibilities for change in the watershed. Mary Stoertz, an Ohio University hydrogeologist whose research laid the groundwork for what would become the MCRP, remembers that 'the agencies [in Appalachian Ohio] did not believe we could clean up Monday Creek. I spent a lot of hours on the phone giving pep talks to guys who were highly doubtful' (Bernard and Young, 1997, p173). Disbelief has started to shift, however, to belief in possibilities, and notable changes have been observed in attitudes of the U.S. Forest Service in the region. Collaboration between MCRP and local agencies like the Forest Service and the Ohio Department of Natural Resources has become the norm (Bernard and Young, 1997, p173). The MCRP received a major grant from the U.S. Environmental Protection Agency in 1996, and Rural Action subsequently received the prestigious National Award for Citizen Involvement for Sustainable Development from Renew America.

Bernard and Young indicate, the MCRP is powered by innovative, pragmatic and flexible people – citizens and professionals – who are willing to collaborate and whose first concern is making the place they call home better (Bernard and Young, 1997, p174). The citizens and professionals of MCRP have developed a watershed management plan and undertaken construction projects, like the capping of the Rock Run gob pile, which have already had a noticeable effect on the flow of acid drainage into the creek. There are now fish in parts of Monday Creek where they have not been found for over 60 years (*Rural Action*, 2001). The success of these initial efforts has led Rural Action to participate in other local watershed restoration efforts. Rural Action has guided the formation of two new restoration groups/projects in the Sunday and Federal Creek watersheds, and is providing VISTA assistance to a third group in the Raccoon Creek watershed. It is hoped that the cutting edge techniques being developed by these restoration efforts will eventually be used by other AMD-affected watersheds throughout Ohio.

As with other Rural Action initiatives, the key to the success of the Sustainable Environments Initiative has been organizing which seeks to involve the community in an inclusive remembering and re-imagining of community. Like the Community Murals project, one important organizing tool has been the monthly potluck dinner. Citizens come together at these meetings with academics and officials to share their memories and hopes for their watershed. In a relaxed environment, common concerns are shared and transformed into meaningful policy through participatory decision-making, so that citizens not only engage in the physical work of restoring the environment, but also in the planning of this work. Increasingly, the arts are being incorporated into the watershed organizing and restoration. The Monday Creek Restoration Project has engaged in a partnership with another organization, Art and AMD, Inc., with funding through a unique collaboration between the U.S. Office of Surface Mining and the National Endowment for the Humanities. Rural Action and MCRP were chosen for this program because of their success in building an action team in the site community, and had demonstrated to government officials that Rural Action could be relied upon to help interpret the project into the community and get input from the community as well. The program involves the use of a landscape architect as a community organizer, engaging community members in a process of designing a community park which also serves as a passive reclamation system for AMD-damaged waters such as Monday Creek. The end result is not only reclamation of a part of the watershed, but also a locus for community interaction and the building of social capital which can be used to expand citizen-led restoration work throughout the watershed. As Bernard and Young indicate, the 'strengthening partnership [of MCRP] gives the dreamers in this watershed hope that Monday Creek will one day not only become a fisher's paradise but also a place whose future residents – the school kids of today – remember the good old days when they helped their watershed heal' (Bernard and Young, 1997, p174).

Conclusion

Organizationally, Rural Action has always tried to hold at its center the tension between organizing and development, believing that the best elements of the group's work will emerge from this kind of grassroots organizing. This can be a real tension in community development work, because when citizens move into the development 'world', they move into dealing with the world of finance, of real estate bankers, of people who suit up and work at agencies such as the Ohio Department of Development. Most grassroots activists and organizers have not spent much time in this world, having to struggle with events and needs in their own backyard, neighborhood, or town. Often, they do not have the knowledge of where to go with issues or needs, or they lack the courage to go and deal with the development world. One of Rural Action's roles has been to bring citizens, academics and government agencies together so that, if a community identifies a problem or need Rural Action will likely know where they will be able to get help. The key task for Rural Action in this process has been finding out what local communities really need, finding where they can go for resources to help them, and most importantly, helping them identify the assets they already have and can build on. This allows citizens to come away from meetings with a positive feeling that change is possible, because Rural Action works to give the ideas and dreams of citizens careful guidance and incubation, helping them to become policy and/or independent projects.

Rural Action has pursued this work within a larger vision of interconnectedness. This applies not only to the biological interconnectedness between all things which is at the heart of the Rural Renewal Strategy, but also to the social and political interconnectedness of organizations and agencies which work within interconnected communities, economies, and ecosystems. Just as the success of the entire ecosystem is based upon diversity and connectedness, so is the success of the political and social ecologies. Things that do not survive or succeed are things that somehow get out of the system. Thus, collaboration between citizens, organizations, and governmental agencies become essential for the health and sustainability of communities and their environments. Therefore, all of Rural Action's work is collaborative in nature, and seeks to enhance the decision-making power of citizens and communities in Appalachian Ohio by actively engaging them in the process of participatory governance.

References

Ajamian, M. (2002) Note to the Authors, Athens, OH, 26 February.

Barber, B. (1998) *A Place for Us: How to Make Society Civil and Democracy Strong*, New York: Hill and Wang.

Bauman, Z. (2001) *Community: Seeking Safety in an Insecure World*, Cambridge: Polity Press.

Bernard, T. and Young, J. (1997) *The Ecology of Hope: Communities Collaborate for Sustainability*, Gabriola Is., B.C.: New Society Publishers.

Berry, W. (1993) 'Out of Your Car, Off Your Horse', in *Sex, Economy, Freedom and Community*, New York: Pantheon Books.

Bingham, M. (1993) 'Stopping the Bulldozers: What Difference Did It Make?', in Fisher, S. (ed). *Fighting Back in Appalachia: Traditions of Resistance and Change*, Philadelphia: Temple University Press.

Couto, R. (1999) *Making Democracy Work Better*, Chapel Hill: University of North Carolina Press.

Daly, H. and Cobb, Jr. J. (1994) *For The Common Good*, Boston: Beacon Press.

Evans, S. and Boyte, H. (1992) *Free Spaces: The Sources of Democratic Change in America*, Chicago: The University of Chicago Press.

Fisher, S. ed. (1993) *Fighting Back in Appalachia: Traditions of Resistance and Change*, Philadelphia: Temple University Press.

Grapevine – Rural Action Sustainable Forestry Newsletter (1999) Community Based Research Conference Links Rural Action With Community Research Network, Fall, 3.

Held, D. (1995) *Democracy and the Global Order: From the Modern State to Cosmopolitan Governance*, Stanford: Stanford University Press.

Imhoff, D. (1996) 'Community Supported Agriculture: Farming with a Face on It', in Mander, J. and Goldsmith, E. (eds) *The Case Against the Global Economy*, San Francisco: Sierra Club Books.

Kittredge, J. (1996) 'Community Supported Agriculture: Rediscovering Community', in Vitek, W. and Jackson, W. (eds) *Rooted in the Land: Essays on Community and Place*, New Haven:Yale University Press.

Kretzmann, J.P. and McKnight, J.L. (1993) *Building Communities From The Inside Out: A Path Toward Finding and Mobilizing a Community's Assets*, Chicago: ACTA Publications.

Lipka, A. and McDaniel, S. (2001) *Community Murals: Handbook & Case Studies A Rural Action Community Toolbox Book*, Trimble, OH: Rural Action.

Putnam, R.D. (1993) *Making Democracy Work: Civic Traditions in Modern Italy*, Princeton: Princeton University Press.

Rural Action (1998) *Building Healthy Communities*, Trimble OH: Rural Action

Rural Action (1999) *Partners in Leadership Development: A Legacy in the Making*, Trimble OH: Rural Action.

Rural Action (2000) *Good Food Direct! 2000 Catalog*, Trimble OH: Rural Action.

Rural Action (2001) *Abstract: Rural Renewal Strategy 2001*, Trimble OH: Rural Action.
Shuman, M.H. (1998) *Going Local: Creating Self-Reliant Communities in a Global Age*, New York: The Free Press.

Chapter 12

Advancing Knowledge and Capacity for Community-led Development in Rural America

Norman Reid and Cornelia Butler Flora

Introduction

The U.S. has not had an explicit rural policy since the Country Life Commission in 1908. Instead, we have assumed that sectoral programs subsidizing agriculture, timber harvesting, mining, or manufacturing would solve the problems of rural development and rural poverty. And when those areas with the highest levels of subsidies were also areas of highest poverty, individual programs of income transfer were implemented to indirectly address problems of under and unequal development in rural places. Place was dealt with only in terms of infrastructure, with roads or housing or water systems or digital connectivity viewed as the magic bullet to offset the disadvantages of distance and dispersion. Only recently has the U.S. moved to place based programs. These programs, like those in Europe, present a different model of development, with drivers from within the community. Participatory community-led development has proved effective in creating jobs, income, and hope (J. Flora *et al*. 1997).

Community-led development, particularly in rural areas, is not simply a matter of money. It is a matter of hope and of participatory processes toward collective goals and toward increased community leadership capacity over time. There is a great deal of evidence that investing in community capacity contributes greatly to community-led development (Kissler *et al*. 1998; Gilat and Blair, 1997). Participation is more than having meetings and presenting decisions. It means "rethinking the underlying roles of, and relationships between, administrators and citizens" (King *et al*. 1998: 317). At its best, community-led development means moving away from paternalism or hopelessness to active collective engagement. A case in point is the Empowerment Zone and Enterprise Community (EZ/EC) program in the U.S., which over time has learned of the critical role of community capacity building as a necessary and sufficient contributor for sustainable community-led development, particularly in areas of high poverty. Community-led development moves beyond citizens stating their needs and government agencies responding. Citizens from diverse situations analyze their situations and discuss alternatives, gathering resources to move toward priority goals from inside and outside the community.

About the EZ/EC Program

One of the most important policy developments in the field of community development in recent years is the enactment in 1993 of the EZ/EC program (Reid 1999).[1] This program, implemented in three 'rounds' in 1994, 1998 and 2001, resulted in the designation of 58 rural communities as Empowerment Zones (EZ) or Enterprise Communities (EC). For the most part, only areas with high poverty rates were eligible to apply.[2] Applications for the program consisted of community-developed strategic plans. The program's benefits include special flexible grants, tax credits, special priority for other grant and loan programs, and technical assistance. From the beginning, the rural component of the EZ/EC Initiative was conceived not merely as the pot of highly flexible funding that it was, but as a tool by which communities mired in long-term poverty could build the capacity to raise themselves permanently to a better existence. The objective, as seen within USDA, was not only to create jobs and improve public services but also to enhance the quality of local decision-making processes and build the local leadership and organizational capacity needed to sustain these re-energized communities beyond the ten-year designation period.

The principal distinguishing feature of the EZ/EC program is that it provides a structured opportunity to change a variety of local conditions, leading to the development – over time – of a community with the capacity to sustain a process of growth and development. Some of the key elements of the process that characterize and differentiate it from most other federal programs are the following:

- It is *long-term* in nature, extending over approximately a decade, rather than a single point in time.
- It emphasizes a *holistic* perspective on community development by insisting that communities address their issues comprehensively, not as a series of isolated issues.
- It requires active *citizen involvement* throughout the life of the development process, in planning, implementing, and evaluating the community's efforts. In particular, low-income and minority citizens and others who are often shut out of community leadership opportunities are expected to be welcomed to take active roles.
- It recognizes that most rural communities are too small and isolated to thrive or be economically competitive in the modern world, and that their limited resources must be extended through the active use of *partnerships* among internal and external organizations.
- The program does not involve goal setting by Washington, but instead, local citizens, based on their *often-unique visions* for the communities they wish to have in the future, set them.

- The development process is meant to be *strategic* and goal-driven, and not a series of independent projects whose linkage and mutually supportive connections is unclear or nonexistent.
- The development process is *planned*, not random or driven by the availability of dollars.
- Strategic plan implementation is *monitored* by establishing performance benchmarks and monitoring progress in achieving them.
- The federal government and the community are in a *collaborative partnership*, not a "master-servant" relationship.
- The program is *flexible* so that each community can pursue its own goals, using its own configuration of resources, and will be aided by flexibility on the part of the federal government.

A key to understanding the EZ/EC program is that it is not a one-time event but a process that continues over time. While the federal-local partnership established by the program extends for a decade, it is the express goal of the program that the processes created during the first decade be sustained over time after the federal government's participation ceases.

There is no assumption that results will flow automatically from the grants, loans, or tax credits offered to the communities; rather, it is understood that real and lasting benefits require significant local buy-in and investments of time and effort. Also, there is no assumption that economic enhancements – though obviously critical – are sufficient in themselves. In essence, the program acknowledges that what is most important in community development is the community, and that the community is much more than a designated territory where development can proceed in a meaningful way simply by taking actions within that territory. Community is, rather, the people who make it up, the structure of their relationships among themselves and with external partners, their skills, attitudes, beliefs and contributions.

The EZ/EC program is, then, a community-led process for local community development. It is, as well, a program that differs greatly from most others in purpose and methods. It places great emphasis on building the knowledge and capacity of citizens and leaders to implement a highly democratic and intelligent process to enhance community well being. It is, in short, a process that requires continual advances in knowledge about how to establish and sustain such community-led development.

Indeed, we found that when communities invested in building community capacity through board training, capacity building among residents, and leadership development – benchmarks most often funded by the EZ/EC funds rather than leveraged funds – progress toward the community's other self-defined goals and their benchmark measures was more often obtained and they were more effective in leveraging the EZ/EC program dollars (Aigner *et al.* 2001).

The EZ/EC Program has been in operation since late 1995, when the first round of EZ/ECs got their implementation processes underway. Over that time, considerable evidence has been accumulated about the rural EZ/EC program. This evidence comes from two principal sources, the Benchmark Management System (BMS)[3] created by USDA Rural Development to manage and monitor the program and a series of field studies carried out by the North Central Regional Center for Rural Development (NCRCRD) at Iowa State University. The remainder of this chapter reports some of the key findings which have emerged, thus far, from this research.

Program accomplishments from the Benchmark Management System

Overall, rural EZ/ECs have compiled an outstanding record of activity and accomplishment. In terms of revenues raised, the 57 Round I and II EZ/ECs – each averaging about 15,000 residents – had raised just over $3.5 billion by June 2002 (Table 12.1), an astonishing amount given that many of these communities had little or no experience with fund raising prior to being designated an EZ or EC.

Table 12.1 Revenues of Rural EZ/ECs by Source, June 2002

Source of Revenues	*Revenues ($ millions)*	*Percent of Total*
Total	3,504.9	100.0
Grant from EZ/EC Designation	203.4	5.8
State Government	653.5	18.6
Non-profit	66.1	1.9
Local or Regional Government	345.6	9.9
Federal Government	1,449.0	41.3
Private Sector	533.2	15.2
Tribal Government	22.6	0.6
Other	231.4	6.6
Per Community Average	60.4	
Leveraging Ratio (ratio of non-EZ/EC grants to EZ/EC grants)	16.2:1	

These funds have been put to work in a wide range of areas that reflect the diversity of their strategic plans (Table 12.2). The emphasis on business activity in most EZ/ECs is reflected in the 953 business loans made and 854 businesses started or attracted to locate in their communities, and over 32,000 jobs created or saved as a result. At the same time, high rates of participation in education programs and programs for youth reflect the priority given to social areas.

Table 12.2 Selected Accomplishments of Rural EZ/ECs, from Designation to April 2002

Program	*Accomplishment Unit of Output*
New/Improved Water & Wastewater Systems	203
New utility hookups	6,061
Business Loans Made	953
Businesses Started or Attracted	854
Education Program Participants	69,608
Youth participating in programs	27,155
Jobs Created or Saved	32,137
Houses Constructed	1,447
Houses Rehabilitated	3,928
New Health Care Facilities	25
New/Improved Recreation & Tourism Facilities	117
Environmental & Natural Resources Projects	39

Overall, EZ/EC activities were achieved primarily by leveraging funds from non-EZ/EC sources. However, there was considerable variation in the rate of leveraging and the number of funding sources across benchmark categories (Table 12.3). Importantly, the benchmark category receiving the least leveraged funding was community capacity; support for institutional capacity is seldom supported by federal programs and – especially in rural areas – is both highly important and difficult to finance. The ECs, which received far smaller EZ/EC grants than the EZs, were far more effective in both achieving their benchmarks and in leveraging funds to do so, reflecting the fact that the EZs received enough funding to finance many projects without partners (Table 12.4).

Table 12.3 Measures of Community Activity in Addressing Goals, July 2000

Benchmark category	*Ave. percent of goal accomplished*	*Average of number funders*	*Average leveraging ratio*	*Average percent of funds from EZ/EC grants*
Community capacity	68.2	2.6	9.5	59.9
Health	82.0	2.7	5.0	42.4
Education	89.9	2.9	17.1	41.1
Public safety and justice	83.4	3.0	5.0	38.6
Children, youth, families	87.6	3.2	45.8	37.9
Business development	88.3	3.0	18.0	33.6
Housing	96.5	3.2	15.6	28.5
Arts, culture, tourism	62.3	4.1	6.9	27.2
Transportation	66.7	3.5	27.0	18.9
Environment	39.7	2.5	103.3	18.2
All functions	80.9	3.1	20.2	34.0

Table 12.4 Funds Received and Leveraging Ratios, Rural EZ/ECs, July 2000 ($ in millions)

Designation Type	*Round I*		*Round II*		*Total*
	EZ (3)	*EC (30)*	*EZ (5)*	*EC (20)*	
EZ/EC grants	71.6	59.2	1.9	0.9	130.8
Total funds	257.4	793.2	44.3	154.7	1,050.6
Ave. per community	85.8	26.4	8.9	7.7	31.8
Leveraging ratio	3.6	13.4	23.4	163.2	12.1

The relative distribution of funding among benchmark categories is shown in Table 12.5. By far the largest amounts of funding went to business development, followed by infrastructure. Housing and education were a distant third and fourth. Business and infrastructure development were also first and second in the amount of EZ/EC grants devoted to them. But capacity building was third among priorities for the use of EZ/EC grants and derived by far the highest percentage of its support from this source, reflecting the difficulty rural communities face in obtaining support for capacity building from other sources. This finding indicates the importance of funding for comprehensive community and economic development in rural areas.

Table 12.5 Measures of Community Activity in Addressing Goals, October 2001

Category	*Funds from EZ/EC grants (dollars in millions)*	*Total Funds Received (dollars in millions)*	*EZ/EC Funds as a Percentage of Total Funds*
Business Development	61.2	947.0	6.5
Infrastructure	19.9	626.2	3.2
Capacity Building	19.2	26.8	71.7
Children, Youth & Families	12.9	70.5	18.3
Education	10.6	165.9	6.4
Housing	10.3	182.2	5.6
Healthcare	7.0	68.1	10.3
Public Safety	5.1	26.2	19.7
Arts, Culture & Tourism	2.1	34.3	6.2
Environment & Natural Resources	1.4	28.2	4.8
Total	149.8	2,175.3	6.9

Evidence about community-building

The NCRCRD has conducted extensive research in the Round I rural EZ/ECs to assess the evolution and impact of the community empowerment process in these communities. A key finding of this research is that the higher the community's investment in capacity, and the higher the level of community resident participation, the more partners were involved and the more outside funding was leveraged (Aigner et al. 2001). The communities not only view investment in community capacity as important, but have discovered that it also pays off. Participatory governance is not just allocated resources and responsibilities to localities. It must also include explicitly building the capacity of the community to work together, plan strategically, as well as budget properly and access outside resources.

When EZ/EC governing boards allocate time, energy, and funds to increase the community's capacity, these efforts appear to pay off with respect to other kinds of benchmarks as well. The number of benchmarks a community designates as "capacity building" in itself is significant. The more the community elaborates individual capacity building benchmarks, the more benchmarks it is likely to have for children, youth and family issues. Among Round I EZ/ECs, the number of capacity building benchmarks correlates significantly and highly with the number of benchmarks for education, public safety and justice, housing, children, youth and family, and indirectly with transportation. Of the $26.8 million EZ/ECs used to build their capacity, the communities put up $19.2 million from their EZ/EC grant funds. In other words, $.72 of every $1.00 spent on capacity building was funded from the EZ/EC grants and only $.28 was leveraged from other sources (Table 12.5).

Yet this investment increased the effectiveness of leveraging for the other non-business development benchmarks.

Investment in community capacity clearly has payoffs. The EZ/EC communities most likely to invest in community capacity building started from high levels of participation in the formulation of the plan and with elected board members from within the EZ/EC census tracts (the areas of highest poverty). This type of grassroots participation in governance, rather than continued elite dominance, characterized the most successful of the Round I EZ/ECs. These communities, in turn, were most successful when they could depend on a strong base of on-the-ground support by community development specialists who are able to provide technical assistance in community processes, leadership and project management skills, and the transfer of best practices from other communities to meet individual, local needs.

Accountability – but to the local community, not just through financial management to meet regulations for state and federal funders – was critical for success. Transparency in the process, including updated web pages, newsletters, and general knowledge of the names, addresses and phone numbers of board members and the judicious use of committees for implementation and oversight allowed community members to understand where the funds for different projects came from and how the allocations were made. This increased bonding social capital within and outside of the EZ/ECs, as it increased the confidence of governing boards in interacting with outside partners and funders.

Lessons about program implementation

Capacity for participatory governance had to be built at the local level (recipients) for the program to be successful. But capacity also had to be at the state and federal level, as an agency which previously had worked from a command/control model of silo programs had to transform itself to work successfully with local empowerment, decision-making, and implementation. USDA's experience in implementing the EZ/EC Program resulted in important lessons about what is required to implement a community empowerment approach. Because of the high levels of community engagement needed to succeed with this approach, traditional program implementation techniques are ineffective; new behaviors are required at both the federal and local levels. We discuss below some of those requirements and the obstacles to meeting them.

Community level lessons

Although the empowerment approach offers major enhancements in communities' control over their futures and opportunities to build sustainable capacity, it is by no means inevitable that local communities are ready or willing to take advantage of them. At the local level, the most critical single factor affecting the success of the empowerment approach is that the community's leadership and citizens understand and accept the empowerment approach, its underlying principles, and the methods that best support it.

USDA's experience with the program suggests that despite the overall outstanding progress of EZs and ECs in creating jobs, leveraging new resources, and achieving other community enhancements, communities vary in the degree to which they comprehend the empowerment concept. Some communities immediately recognized its importance, began to apply it during their strategic planning processes and have continued to benefit from it during the implementation phase. Others appeared to give lip service only to the principles in making their applications and then neglected or abandoned them altogether after they received an EZ or EC designation.

We find that communities that comprehend empowerment principles have experienced greater benefits from the program than those that do not. This appears to be true not only in the acquisition and use of resources but also in the more intangible aspects of community participation, vitalization, innovation, commitment, and satisfaction.

Among communities that failed to grasp the importance of empowerment principles, two factors appear most frequently. One is clearly related to how local leaders perceive the program. A number of communities, especially in Round I, tended to regard the program as 'a grant' rather than as a community-building process. The resulting attitude could be characterized as 'just give us the money and go away'. Communities that took this position tended to display lack of recognition of empowerment's benefits or willingness to seek to achieve those benefits. Their designation as an EZ or an EC was assumed to be part of the old governance model: resources came from the federal government, probably based on the connections of a few political 'players' in the community, who then continued to do as they always did.

A second factor is the propensity of leaders in some communities to approach the role of leadership through control – of objectives, information, resources, decisions, and participation. This style of leadership often appeared to be more clearly focused on the leader as an individual and his or her recognition and accomplishments and might be termed 'ego-driven' leadership (McAlpine, 2000). On the other hand, leadership that focuses principally on the success of citizens and other participants by enabling and supporting them might be termed 'servant leadership' (Greenleaf, *et al*. 1996). Ego-driven leadership fails to create and foster an open process of active civic engagement that more properly characterizes an empowered community.

Examples of each of these types occurred among the rural EZ/ECs, and they are instructive regarding the importance of leadership style in building the very community capacity that is so critical to overall community development and enhancement. In one case, an Empowerment Zone declared that its "strategy" was to dispense its EZ/EC grant funds through a series of funding competitions that were open to all comers. This community, which was among the most poverty-stricken, did not effectively engage citizen participation in a planning process. Thus no clear priorities were established; no strategic plan implementation unfolded, nor did active community and organizational participation address local issues. Instead, priorities were set *de facto* by the range of applications it ultimately funded. In effect, the 'strategy' was to use the EZ/EC grant funds as a $40 million pot of capital into which many hands could dip. No special effort was made to build community-level capacity to sustain a process of development beyond its ten-year designation or to see that the projects funded added up to a coherent whole.

USDA program officials conducted an extensive series of trainings and negotiations with this Zone to develop specific strategic benchmarks (i.e., objectives and performance measures) that reflected the community's development goals. While initially promising, this effort ultimately had little effect. This community was unable to perceive the Benchmark Management System as a means of managing its strategic planning process, and USDA had to force the EZ to maintain some degree of timeliness in its record keeping and reporting. Not surprisingly, the projects that were funded seldom included partnerships, and their sponsors often used them to bolster the current resources of specific institutions rather than invest in sustainable activities. As a result the community's leveraging ratio was well below the average for all EZ/ECs. It appears likely that many of this EZ's projects will not be sustainable over the long term, a tragic consequence for this community's prospects of emerging from its oppressive poverty.

The second example is a community whose leadership was very heavily ego-driven. Communities that reflect the democratic principles of empowerment set priorities and conduct implementation through open community participation in decision-making processes. Most rural EZ/ECs have striven to follow this methodology, but a few have not, falling back on traditional politics to conduct their business. The most extreme instance involves a community that has been under the direction of a single individual for a number of years. Exemplifying what we call ego-driven leadership, the community's entire application process was organized by this person, who sought the EC designation both as 'a grant' and for the high-visibility national recognition it conferred. Immediately following the EC's designation, the official attempted to eliminate the portion of the designated area that lay outside his personal political base. He attempted to dominate the local EC board with handpicked representatives and was stopped only by action from USDA officials. Not surprisingly, the community has been one of the most resistant to employing the BMS for plan management, choosing instead to view it as a bureaucratic obstacle. To date, this community's performance is among the poorest, with little money expended, few jobs created and an extremely low leveraging ratio.

Happily, we can report a larger number of counter illustrations. The third case is a community whose application was initially written by a local organization in the name of the larger community comprising the EC area. Following designation, the organization planned to keep the EZ/EC funds for itself and to control the selection of projects and their implementation. A period of citizen complaints ensued, however, and citizens of the community organized to take control of the EC designation by electing a new governing board and engaging a different local organization to carry out executive functions. Though its first few years were rocky, this community ultimately put together a solid, community-oriented program that drew on underutilized local assets and addressed the area's unique needs. This community has created or saved many jobs and compiled a high leveraging ratio. This example is by no means unique; similar citizen-led efforts to assure broad-based citizen control have taken place in several other rural EZ/ECs. In addition, many more EZ/ECs – especially those selected in the second round of competition – began with a sound understanding of the empowerment process from the beginning so that fewer leadership adjustments have been necessary.

National level lessons

The EZ/EC program had to be developed within a very short period of time, too short to put in place – or even to understand – all of the requirements for making it successful. Congress passed the legislation that established the program in August 1993 with a deadline of December 31, 1994 for making the first round of designations. From this 17 month period thus established by law, six months were reserved to review the applications, and as much time as possible for applicant communities to develop strategic plans – in the end only four and a half months, a period subsequently judged insufficient for the serious community development work needed. In the remaining six and a half months, a new approach to rural development had to be conceptualized, regulations written and published, an organization established, staff trained, information materials written, and regional training conferences for communities conducted. Under these circumstances it is understandable that the implementation of the first round of the program revealed that implementing participatory governance as a tool for poverty reduction required many unanticipated capacities from both the local communities and the Federal government. The lessons were many. Here we focus on three major areas

- The paradigm underlying program goals and procedures;
- Performance measurement; and
- Information and technical support.

Shifting the paradigm

The empowerment process reverses the direction of responsibility for development from what has been usual in the past. Rather than building projects, the program's objective is to build communities with the capacity to do things for themselves. Community capacity is therefore regarded as more important in the long run than any number of programs to increase business activity or meet local infrastructure or other needs. But this concept of development is hard for many leaders to understand or accept, and the popular wisdom that drives most rural development policy rests on much simpler notions of what constitutes rural development. The empowerment program also requires national leadership that understands the underlying principles of empowerment and the kinds of support needed to help it succeed. Willingness to set broad parameters for the program, enforce their integrity and provide information and technical assistance while leaving the details of implementation to local communities are key elements of the new leadership paradigm embedded in the participatory governance model. In short, the same kind of servant leadership is needed at the national level as at the community level.

In practice, however, this type of leadership support has been difficult to achieve and sustain. In its early years, the EZ/EC program was supported vigorously as a special presidential initiative by political appointees who had little understanding of the program's essential nature or special needs. While their connections enabled them to overcome many short-term obstacles, they contributed little to building the organizational infrastructure needed to sustain and enhance the program. Furthermore, some of these officials treated the program as just another opportunity to make favorable public appearances and created costly misunderstandings by carrying inappropriate messages to communities. In the end, by treating the program as a special initiative they placed continued support for the program at risk of automatic opposition by succeeding administrations based on the perennial view that the previous administration could do no right.

In the U.S., rural development is broadly mistaken by politicians as a field that does not require special knowledge and by civil servants as something that can be achieved piecemeal through individual, stovepipe programs. The empowerment program rests on entirely contradictory premises—that appropriate development relies on goal-directed strategies that are holistic and interconnected, based on solid knowledge and informed choices.

Over time, it became clear that the EZ/EC program had to make the transition from special initiative to program, with accompanying attention to all the mundane but essential matters of organization and follow-through. It also became clear that policy officials had to be educated about the differences in paradigm required to assure the program's success, a task often outside the capability of career program managers to achieve. Finally, the need to protect the program during a change in administrations led to efforts to detach it from the administration that originated it and consolidate it as a regular – albeit unique – program. It would be premature to conclude that – in the context of a national bureaucracy long accustomed to project-by-project largesse as a method of promoting community development – the paradigm shift has been successfully achieved and participatory governance established.

Measurement systems

Because the EZ/EC program is so novel and outside the experience of a political system that measures success based on dollars to districts, it was clear from the beginning that objective measurements would be needed to demonstrate its effectiveness. Two sets of actions were called for: to monitor community progress and to measure long-term impacts. From the start, the Office of Community Development (OCD) collected information about activities underway in EZ/ECs. At first, communities reported on paper, but it was soon obvious that the volume of data would overwhelm OCD's capacity to make meaningful use of it. OCD turned to collecting information over the World Wide Web and developed the Benchmark Management System (BMS). The basic principle behind the BMS was to give EZ/ECs a tool to track their own project status, funds status, and accomplishments for regular reporting to their governing boards and citizens. At the same time, the BMS system served as a workflow system for USDA state offices to review and approve strategic plan amendments and narrative progress reports. The data were collected in a national database, which gave OCD up-to-date information for program assessments and reporting. These data have demonstrated the extent of community progress to policy officials eager for immediate results from this long-term program. The fact that this progress has been so incredible has demonstrated beyond doubt the program's effectiveness in mobilizing community activity.

In addition to short term improvements in community services and economic activity, the EZ/EC program is intended to produce long-term strengthening of civil society and local capacity to sustain development. Thus, it is also important to measure change in the underlying capacity within the communities. At the very beginning of Round I, baseline data were gathered for the census tracts making up the three EZs and 30 ECs. In addition, narrative case studies described the state of affairs at the time the program began. Subsequently, social indicators were gathered for Round I EZ/ECs and analyzed by NCRCRD in conjunction with data indicating the extent to which EZ/ECs had incorporated practices reflecting civil society; some of these results are reported in this chapter.

At this point in implementing the EZ/EC program, we can only draw preliminary conclusions about the program's ability to bring about lasting changes. In fact, even Round I is still in mid-course, and many of the fundamental changes in community organization and behavior sought will develop only over the long term. It is unrealistic to expect a vibrant participatory governance model to be fully developed at this point in time. Therefore, the measures of activity levels, resource acquisition and use, and outputs from project investments are the most valid measures of program accomplishments currently available.

Nonetheless, the research reported here shows that how a community is organized for civic participation, decision-making and other indicators of civil society affects its levels of activity and accomplishment. In addition, higher levels of civil society are very likely correlated with empowerment within the community. And it is empowerment that appears to enhance sustainability of the community's efforts and improvements in well-being. An important learning from implementation, therefore, is that measuring and understanding the internal dynamics of the program, and not merely its activity levels, are essential in order to adjust program methods in mid-course. Nowhere is this truer than in a complex and paradigm-shifting program such as the EZ/EC program.[4]

Information and technical support

Because the EZ/EC program required a paradigm shift for communities as well as the federal government, and because it centered on a process of strategic planning to better position communities to thrive economically and socially, information, knowledge and technical skills are clearly critical assets. This is especially true for rural areas, where leaders are often ordinary citizens working part-time on a voluntary basis and whose determination may exceed their training and access to needed information. In addition, the empowerment process requires communities to undertake activities that may be unfamiliar or even uncomfortable. Cooperating across rival towns or counties and sharing power by organizing into task forces or committees are but two examples.

Helping rural communities understand strategic planning, methods for establishing meaningful performance benchmarks and how to access resources and implement projects effectively are very staff-intensive processes. Within the Rural Development (RD) mission area of USDA, of which OCD is a part, community development technical assistance has traditionally been defined as 'everybody's job', which often means it is adjunct to other, more central functions and not a measurable part of employee performance. In addition, it was often assumed that the most needed skills were the ability to 'meet and greet' and that the objective was 'selling' RD's funding programs. Few staff members were trained specifically in community development methods, and in many parts of the organization no clear structure for internal support for the community development function existed.

In 1996, two years into the implementation of the EZ/EC program, 167 positions in field offices were assigned to support the EZ/EC program. However, allocation of these positions was left to the discretion of state RD offices, and many were directed to other purposes or assigned to the EZ/EC program in name only. In 2002, a survey of staff with community development functions showed that the number of reported staff years devoted to the EZ/EC program had shrunk, even though the program had nearly doubled from 23 to 58 communities during that time. Although technical assistance is a critical input that differentiates the EZ/EC program from other programs, the willingness of policymakers to invest personnel assets in this manner led to a sharp decline in the levels of support available.[5]

Another concern that emerged during the course of EZ/EC program implementation is the need to expand the capacities of field staff to provide effective technical support. OCD responded in several ways: by holding skill-training conferences, providing special training opportunities, and developing materials and applications on the Internet to support both community development staff and the communities they serve.

Training sessions for field staff began almost immediately following the program's enactment in 1993 and continued in the early years of the program, along with joint rural-urban conferences shared with the Department of Housing and Urban Development, which administered the urban portion of the EZ/EC program. In 1997, OCD began an annual program of rural community development training conferences that included both RD field staff and representatives from EZ/ECs. Over time, these training conferences have shifted emphasis from understanding the requirements of the EZ/EC program to tools and techniques for fostering and sustaining empowerment within the community. In addition, OCD sponsored a series of weeklong advanced rural development training for field staff and EZ/EC leaders to build broad knowledge of rural development issues. Although Rural Development has neither required nor sponsored professional certification in economic or community development, a few staff have obtained this on their own, and OCD is exploring ways in which certification can be achieved for other staff and community leaders.

An important lesson from implementing the EZ/EC program is the need to document technical assistance. OCD is presently developing a system to measure the types and amounts of technical assistance provided, the outcomes, and an on-line toolbox of materials that technical assistance providers can draw upon. The objective is to document what is now considered ephemeral, as well as its impact, and to enable even partially trained staff to provide proficient assistance to rural communities. Participatory governance does not mean that localities no longer need technical expertise. But it does mean that they know the right questions to ask to be sure that the technical assistance is most helpful in reaching strategic goals.

Conclusions

USDA's experience in implementing the EZ/EC program has resulted in many lessons. Clearly, one set relates to the value of approaching rural development by using a holistic, long-term and inclusive process and honoring broad-based local definitions of development. Where this process has been most clearly adhered to, the best outcomes appear to have occurred. Another is that building community capacity is important in affecting community-level outcomes, rates of success and the probability that these gains can be made sustainable.

Investment in community leadership capacity pays off, and ensuring widespread participation makes a difference. To foster capacity and participation, OCD developed information support systems for its own staff, changed program procedures to better implement empowerment principles, and invested in its own capacity to work with rural community partnerships and grassroots governing structures. As a result, OCD was able to apply the lessons learned from the Round I communities in structuring the Round II and Round III competitive processes and selection procedures and by increasing the amount of community capacity building and board training collectively available.

The program's specific instructions that not only required participation but also specified the mechanisms to ensure that it was broad-based and open were critical to the program's success. Those EZs and ECs that paid minimum attention to broad-based strategic participation in the formulation of the proposal and in naming the project leadership achieved the least.

Another important conclusion is that money is not the main ingredient for sustainable community-led development, although the opportunity to access it spurred participatory community strategic visioning and planning. NCRCRD analysis found that the amount of automatic EZ/EC grants provided to the communities exerts a powerful negative influence on the community's efforts to raise funds from other sources. In both Round I and Round II, the leveraging ratio was greater for the ECs than the EZs.

When the program was viewed as a grant, rather than a process, the result was limited impact. To the extent communities treated the EZ/EC opportunity as a strategic process to create partnerships to reach goals determined through participatory processes, EZ/EC communities conformed to the key principle of community-based partnerships, developed sustainable organizational forms through broad participation, and created economic opportunities. Blair (forthcoming) found in Nebraska that when mechanisms for planning were clearly described and facilitated, plans were more likely to be implemented; this appears to be the case for EZ/ECs as well.

A wide range of research supports the importance in investing in community capacity (Gasteyer *et al.* 2002; Flora and Luther 2000; Luther and Flora 2000; Flora *et al.* 1996). And when that investment is tied to other resources and continuing support, measurable impacts on community well being occur. But for participatory governance to occur, capacity must be built at both the local and the federal levels.

References

Aigner, S.C., Flora and J. Hernandez (2001) 'Empowering Sustainable Communities: The Premise and Promise of the Empowerment Zone/Enterprise Community Initiative', *Sociological Inquiry*, 71(4), Fall.

Aigner, S., V. Raymond and S. Tirmizi (2001) *Empowering Rural Communities: A Perspective After Five-Years*, Ames, IA: North Central Regional Center for Rural Development, http://www.ncrcrd.iastate.edu

Blair, R. (forthcoming) 'Public Participation and Community Development: The Role of Strategic Planning', *Public Administration Quarterly*.

Flora, C. and V. Luther (2000) 'An Introduction to Building Community Capacity', in P.V. Schaeffer and S. Loveridge (eds) *Small Town and Rural Economic Development: A Case Studies Approach*, Westport, Connecticut: Praeger, pp. 1-3.

Flora, C.J. Flora and K. Wade (1996) 'Measuring Success and Empowerment', in N. Waltzer (ed.) *Community Visioning Programs: Practices and Principles*, Westport, CT: Greenwood Press, pp. 57-74.

Flora, J., J. Sharp, C. Flora and B. Newlon (1997) 'Entrepreneurial Social Infrastructure and Locally-Initiated Economic Development', *Sociological Quarterly*, 38 (4): 623-645.

Gasteyer, S., C. Flora, E. Fernandez-Baca, D. Banerji, S. Bastian, S. Aleman, M. Kroma and A. Mears (2002) 'Community Participation for Conservation and Development of Natural Resources: A Summary of Literature and Report of Research Findings', *Delta Development Journal*, 1 (2), 57-75.

Gilat, O. and R. Blair (1997) *Strategic Planning: The Nebraska Way*, Lincoln, NB: Partnership for Rural Nebraska.

Greenleaf, R., D. Frick and L. Spears (eds) (1996) *On Becoming a Servant-Leader*, San Francisco: Jossey-Bass.

King, C. K. Feltey and B. Susel (1998) 'The Question of Participation: Toward Authentic Public Participation in Public Administration', *Public Administration Review*, 58 (4):353-359.

Kissler, G., K. Fore, W. Jacobson, W. Kittredge and S. Stewart (1998) 'State Strategic Planning: Suggestions from the Oregon Experience', *Public Administration Review*, 58 (4):56-82.

Luther, V. and C. Flora (2000) 'Capacity Building and Leadership in Yuma, Colorado', in P. Schaeffer and S. Loveridge (eds) *Small Town and Rural Economic Development: A Case Studies Approach*, Westport, CN: Greenwood Publishing Group.

McAlpine, A. (ed.) (2000) *The Ruthless Leader: Three Classics of Strategy and Power*, Chichester, NY: Wiley.

Reid, J. (1999) 'Community Empowerment: A New Approach for Rural Development', *Rural Development Perspectives*, 14 (1): 9-13.

Wang, F and J. Van Loo (1998) 'Citizen Participation in the North Delta Mississippi Community Development Block Grants, Empowerment Zones and Enterprise Communities', *Planning Practice and Research*, 13 (4):443-452.

Notes

1 The EZ/EC program was established to operate in both rural and urban areas. The Office of Community Development within USDA Rural Development implemented the rural program; the urban program was implemented by the Department of Housing and Urban Development. Although the same statute governs both rural and urban EZ/ECs, differences in community development aspects emerged during implementation. This chapter focuses exclusively on the rural program.

2 Beginning with Round II in 1998, areas whose distress was based on population outmigration became eligible for a limited number of designations.

3 The Benchmark Management System (BMS) is a Web-based strategic plan management system developed by USDA. The BMS, winner of the 2000 eGov Pioneer Award, is used at the community level to manage strategic plan goals, activities, resources, workplans, and accomplishments; at the USDA Rural Development State Office level to monitor strategic plan amendments and progress; and by the USDA National Office to manage overall program performance and monitor program achievements.

4 As a result, program implementation actions have been taken to enhance these community processes, especially during the implementation of Round II and in revised program regulations issued in final form in 2002.

5 It is important to note that the EZ/EC program was developed concurrently with two trends that made such staff investments increasingly difficult – a sharp rise in the number of grant and loan programs and the levels of appropriations for RD programs overall during the last decade, and a sharp reduction in the number of field offices and staff within Rural Development. Nonetheless, the staff cuts were disproportionately taken from community development technical assistance.

Chapter 13

Managing Partners: Performance Based Contracting in Baltimore's Empowerment Zone

Robert Stoker

Introduction

This chapter deals with the management of inter-governmental partnerships through performance-based contracting. In 1994 Baltimore was designated as one of the six original urban empowerment zones. Empowerment zone designation provided a variety of tools to address local problems and exploit local opportunities for urban economic development, especially job creation. Like the other empowerment zones, Baltimore has a partnership relationship with the Department of Housing and Urban Development (HUD) that is managed by a performance-based contract. However, the emphasis the empowerment zone initiative placed upon the development of unique, local plans created a difficulty for partnership management – how to direct and evaluate partners in a process in which the means and ends of the relationship are uncertain.

The theoretical material I describe in the chapter is illustrated with a variety of examples from Baltimore, Maryland's Empowerment Zone. Material from Baltimore demonstrates the ambiguity inherent in performance monitoring, the need for flexibility in the development of implementation plans and performance benchmarks, and the limitations of control that undermine attempts to hold partners responsible for performance. Case material from Baltimore also illustrates that performance-based contracting is an incomplete strategy to manage partnerships when programs are complex and multi-faceted.

The uncertainty inherent in the empowerment zone initiative required extraordinary flexibility on the part of HUD. Initially HUD managed their local partners by allowing them to develop unique implementation plans and performance benchmarks reflecting key differences in local programs. However, in 1998 HUD created an automated performance monitoring system that was designed to facilitate cross-site comparisons and reporting in plans and benchmarks and reduced the effectiveness of performance-based contracting as a means to manage the partnership relationships. If its limitations and potential are appreciated, performance-based contracting can play a constructive role in the management of partnerships.

Managing partners

Although intergovernmental partnerships can improve public sector performance (Radin *et al.*, 1996), managing partners is a challenging problem. Partnerships introduce conflict and uncertainty into the implementation process and, even when partners agree about policy goals and objectives, cooperation can be difficult to achieve (Stoker, 1991). To overcome these difficulties program sponsors and partners must bargain to establish the terms and conditions under which they will work together. One useful framework for managing intergovernmental partnerships is performance-based contracting because, when properly managed, such contracts encourage cooperation. This chapter examines performance-based contracting and describes an illustration – the Performance Monitoring System (PERMS) used by the Department of Housing and Urban Development in U.S. Empowerment Zones, and in the Baltimore, Maryland site in particular.

Daniel Elazar defines partnership as coordination 'among several centers that must negotiate cooperative arrangements with one another in order to achieve common goals' (1972, p.3). These arrangements are often expressed as a contract (or sometimes less formally as a memorandum of understanding). However, negotiating the terms under which intergovernmental partnerships are managed empowers implementation participants and affords them the opportunity to bend policy toward their own concerns and priorities. If care is not taken partnerships can become a means for implementers to capture policy and move it away from its original purpose. However, conflict about the means or ends of policy need not be disabling; partnerships can be crafted to encourage cooperation despite conflicts of interest (Stoker, 1991).

Performance-based contracting encourages cooperation in two ways. First, it enlarges 'the shadow of the future' (Axelrod, 1984); participants make decisions in the present in anticipation of their consequences for the future. This is accomplished by linking performance in the present to future decisions about financial support. Second, performance-based contracting helps participants clarify their expectations of one another by making the terms governing the relationship transparent to all (Oye, 1985). Specifying what is to be done by whom and how performance will be evaluated creates transparency and encourages cooperation by easing uncertainty about the consequences of partnership. However, intergovernmental partnerships are a special case of performance-based contracting. When such contracts are used to manage partners, the purpose of the contract, the political dynamics of the relationship between sponsor and contractor, and the contracting process are altered.

Performance-based contracting

Performance-based contracting is a system to govern the relationship between sponsors and contractors. The contract specifies what is to be accomplished by whom; establishes indicators to monitor performance to assure that the specified work is being done; and distributes rewards or punishment on the basis of performance. However, the details of the contract management process depend upon the type of relationship sponsors want to establish with contractors. If contractors are suppliers, the sponsor designs the contracting process to specify the contractor's responsibilities in advance and clarify the consequences of poor performance (usually termination of the contract). If contractors are partners, the sponsor must be more flexible about the contractor's responsibilities and designs the contracting process to encourage transparency so that performance monitoring can detect and correct problems through technical assistance (contract termination is undesirable when contracting with partners and is used only as a last resort).

Performance measurement is the foundation of performance-based contracting. Harry Hatry defines performance measurement as '*measurement on a regular basis of the results (outcomes) and efficiency of services or programs*' (italics in original) and links performance measurement to the effectiveness of public policy through the process of managing-for-results, 'a customer-oriented process that focuses on maximizing benefits and minimizing negative consequences for customers of services and programs' (1999, p.3). He describes a plan for implementing performance measurement including the following elements:

- The mission, objectives, and customers of the agency or program must be identified.
- The results (or outcomes) the program seeks to achieve must be identified.
- Indicators of the outcomes must be identified (usually in consultation with stakeholders).
- Data sources for the indicators must be identified or developed.
- Data collection procedures must be devised and implemented.
- Benchmarks to which the indicators are compared must be developed.
- An analysis plan for the performance data must be devised.
- The role of partners in the performance measurement process must be determined.

Managing suppliers with performance-based contracts

Performance-based contracting disciplines and directs suppliers by linking their performance to financial rewards. Hatry (1999, pp. 170-171) notes that performance indicators and benchmarks must be carefully developed and included in contract or grant agreements. He suggests that contract sponsors include performance indicators in requests for proposals (RFPs). Benchmarks also may be included in RFPs or negotiated later as part of the contractual agreement. Rewards and penalties are included in these agreements including contract termination for non-performance. The threat to terminate funding, though rarely used, is a key source of discipline. The greatest power the sponsor has is to break off the relationship with one contractor and search for another.

When managing suppliers, performance-based contracting is done well when the sponsor's guidelines are clear, detailed, and unambiguous and the link between performance and financial support is enforced rigidly. The sponsor may specify the product or service to be provided in detail or simply specify the desired outcomes, leaving to contractors the decision about how best to accomplish them. The sponsor requires and receives detailed responses that commit contractors to specific actions or accomplishments in exchange for financial support; this settles the contractor's responsibilities in advance of making a commitment to fund the project. After the award the sponsor must closely monitor the contractor's activities and accomplishments according to the framework established during negotiation of the agreement. The sponsor must then follow through by rewarding or punishing the contractor on the basis of performance. The positive effects of performance-based contracting can be seen in the management of routine procurement contracts.[1] Contractors who supply quality goods or services, on schedule, at the right price are rewarded with additional business. Those who provide shoddy goods or poor services, miss deadlines, or overcharge lose out.

Although contracts protect suppliers from the whim of sponsors by specifying the terms of exchange in detail, the real purpose of performance-based contracting is to help sponsors hold contractors responsible for results. If the contractor is not competent, he is replaced. The sponsor does not invest in technical assistance to improve the contractor's capacity to perform. The contractor understands that he must produce what the contract specifies or suffer loss of business (or financial support). Ultimately, this is expected to improve public sector performance. Sponsors are attracted to competent contractors with established track records and positive performance reviews. Sponsors are empowered as consumers; they shop for suppliers and form relationships with capable contractors for as long as their services or products are needed.

The contracting process described above reflects several key assumptions about conditions that make contracting-out an appropriate public management strategy. These conditions include: that there is an active, competitive market to provide the good or service; that the government has a clear vision of the product or service for which it is contracting; and that performance-monitoring problems are manageable (Kettl, 1993). This contracting process is based upon the supposition that sponsors enjoy supply-side competition; there are numerous contractors eager to receive financial support in exchange for providing some good or service to the sponsor (or to others on the sponsor's behalf). The contracting process also requires a clear, fixed vision of what is to be accomplished. The vision, which may be developed in consultation with contractors, may be a detailed plan or simply a statement of the desired outcome. However, the vision must be clear and stable before performance indicators and benchmarks can be established. For performance measurement to be feasible, technical and political difficulties must be overcome. The technical difficulty is to develop measures that truly indicate the quality of public goods or services provided (Newcomer and Scheirer, 2001). The political difficulty (which is related to the technical difficulty) is to gain cooperation from contractors in a process that is used to reward or punish them. All of these key assumptions about the contracting process change when performance-based contracting is used to manage intergovernmental partnerships.

Managing partners with performance-based contracting

Partnerships are a special case for performance-based contracting (Hatry, 1999). Partnerships are typically formed when the means to deliver services or produce goods are controlled by a few key organizations (typically state or local governments or highly specialized firms). As a result, supply-side competition is limited. Certainly, the selection of potential partners is limited when forming intergovernmental partnerships. This is an important deviation from the assumptions of performance-based contracting that is used to manage suppliers. No longer is the sponsor a consumer shopping in a competitive marketplace. When forming partnerships the sponsor is likely to target specific governments or firms as partners because of the distinctive contributions these potential partners can make to the project (Stone, 1989).

Partnership changes the purpose of the contract and the nature of the relationship between sponsor and contractor. Outside of partnership, the purpose of performance-based contracting is to hold suppliers responsible for results. The relationship is essentially economic; the sponsor designs the contracting system to expedite exit from the relationship if the supplier's performance is not satisfactory (Hirschman, 1970). However, it is self-defeating for a sponsor to exit from partnership and search for another contractor. If partners contribute something special to the project the sponsor cannot 'improve performance' by abandoning the relationship. Within partnerships, the purpose of contracting is to bind the parties in a committed relationship so they can develop common goals and cooperate to achieve them. In this context, exit is painful for the sponsor and is used only as a last resort. The relationship between sponsors and partners is more political than economic (Hirschman, 1970).

In partnerships, transparency and clarity of the expectations among participants is more significant to gaining cooperation than the threat to withdraw financial support. Partners and sponsors work together to detect and correct problems during the implementation process. Sponsors depend upon the good faith of partners to implement performance measurement because it is an ambiguous, expensive, complex, and time-consuming task. If partners perceive performance measurement as a threat to their financial support, they are unlikely to cooperate in monitoring progress and reporting difficulties. To the contrary, when threatened, partners are likely to view performance monitoring as a "public relations" problem and strive to present a good face to the sponsor, compromising the value of information the sponsor receives. To avoid this difficulty, the sponsor must negotiate and manage the contract to detect and correct problems in cooperation with the partner. Rather than using performance indicators to distinguish capable contractors from poor ones (as one does when contracting with suppliers), the sponsor links performance monitoring to technical assistance programs aimed at improving the partner's performance.

Hatry (1999) summarizes these changes and suggests that it is prudent to encourage participation by partners in all aspects of performance measurement: '*Preferably, programs would work with other organizations as partners in designing and implementing the whole performance measurement process*' (p. 46, italics in original). Subsequently, he identifies six steps to managing 'intergovernmental performance partnership' paraphrased below (pp. 170-171).

- The central agency and representatives of local agencies jointly select a set of outcome indicators.
- The local agencies provide quarterly information on each data element to the central agency.
- The central agency tabulates the data for each agency.
- The central agency provides each local agency with summary data for each outcome indicator and with comparisons among local agencies.
- Technical assistance is provided to poorly performing agencies.
- The central agency and local agencies sponsor an effort to identify and disseminate information about exemplary practices.

These changes amount to a significant revision of the contracting process. However, within partnerships performance-based contracting must be even more flexible and dynamic than these changes imply. Contracts are based upon plans that contain assumptions about organizational capacity, context, and target group responses to programs. These assumptions may be inaccurate or incomplete. Organizational capacity may not be what was expected because the implementation process exposes the hidden costs and difficulties inherent in the policy (Stone, 1980). Context may change; especially if the initiative is implemented over a long period of time or other policy interventions occur. Members of the target group may not respond as expected, especially if programs are innovative. If so, it is uncertain whether the plans that are the foundation of the contract will work as expected and this introduces uncertainty into the relationship between sponsors and contractors.

Uncertainty does not make performance-based contracting impossible, but it does require additional flexibility in contract management. Unlike suppliers, partners expect to influence the means and ends of the project. Initially, the sponsor's vision must be broad and flexible to allow negotiation with partners about the specific meaning and purpose of the project so that the entire project, from the definition of the problem to the possible solutions, reflects the partner's perspective. This infuses the project with knowledge about local context and capabilities that sponsors lack. Beyond this, uncertainty about the assumptions behind policy implies that it may be prudent to revise policy during the implementation process as knowledge and experience are gained. A mid-course adjustment is problematic, however, if the sponsor's concern is to discipline the contractor by specifying performance expectations in advance. Sponsors and partners must agree to continuously monitor progress, but also to be open to re-negotiate their plans and expectations as knowledge is gained or circumstances change. As a result, performance monitoring becomes an iterative process that is open to reconsideration as new developments emerge.

Performance-based contracting in the Empowerment Zones

Performance-based contracts are used to manage intergovernmental partnerships in the Empowerment Zone/Enterprise Community (EZ/EC) initiative.[2] The program is widely noted for its innovative use of regulatory relief, tax incentives, and wage subsidies to encourage employment and economic development. Less noticed, but equally important are the intergovernmental partnerships at the heart of the EZ/EC initiative. These partnerships are governed by agreements between the federal government and local partners that include the development and operation of a performance monitoring system.

The EZ/EC initiative emphasized the importance of mobilizing local communities to develop solutions to the problems of economic renewal and community revitalization.

> The Initiative recognizes that local communities, working together, can best identify and develop local solutions to the problems they face…The program provides performance-oriented, flexible Federal grants so communities can design local solutions that empower residents to participate in the revitalization of their neighborhoods. (HUD, 2001a)

The emphasis the initiative placed upon local solutions is consistent with partnership because it features a broad, flexible vision that allows partners to have significant influence over the project's specific meaning and content.

The federal government conducted a competition for designation. One hundred and five communities were selected as Empowerment Zones or Enterprise Communities in the initial process completed in 1994; six urban Empowerment Zones and three rural Empowerment Zones were designated (HUD, 2001a). Urban Empowerment Zones are geographic areas within cities targeted for renewal through the use of a variety of tools including $100 million social services block grant, tax incentives, regulatory relief, and technical assistance. HUD was designated as the lead federal agency for the urban sites. One of the urban Empowerment Zones selected was within Baltimore, Maryland.

The selection process organized by HUD reflected the emphasis the EZ/EC initiative placed upon local problem solving. Potential sites (local communities that completed an application for EZ/EC designation) were directed to fashion a strategic plan that built upon their unique strengths while also reflecting local needs and priorities. The process encouraged community mobilization; sites were encouraged to involve numerous, diverse elements of their local communities in the development of their strategic plans. This allowed sites to articulate different visions of the purpose of the initiative to reflect local context, not just to propose different means to accomplish an established purpose. The strategic plans submitted by the applicants included objectives, programs to accomplish the objectives, and performance benchmarks that could be used to assess progress.

Following selection, the designated sites were required to make additional submissions on benchmarks that further elaborated and explained their plans and how progress could be measured. This became the foundation of the performance monitoring system HUD put in place to facilitate the reporting and monitoring necessary to support the partnerships. HUD enforced its demand for additional submissions by withholding funding from the selected sites until they were completed. By linking designation to performance monitoring, performance-based contracting became an integral part of the EZ/EC initiative. However, HUD's decision to withhold funds until sites complied with their demand for additional submissions limited community participation in the benchmarking process in Baltimore. Although the community had been involved extensively in developing the strategic plan and HUD wanted to encourage community participation in the implementation process, the community's role in the benchmarking process was limited because officials in Baltimore wanted to expedite the submission of benchmarks in order to qualify for funding. Involving the community invites delays, so the submission of benchmarks was handled by professional staff that city agencies and philanthropic organizations had detailed to the city's empowerment zone initiative to assist with the planning process.

Several aspects of EZ/EC program administration reflect the flexibility required to manage intergovernmental partners. HUD was flexible about the meaning and purpose of the initiative and respected local representatives as partners in program development. Each of the six urban Empowerment Zone sites had different objectives and programs to achieve them; consequently, each site had unique performance monitoring requirements. This flexibility extended to changes made during the implementation process. New programs were developed as opportunities came up; old programs were dropped or revised. HUD made an unusual financial commitment to its partners; funding was provided for a ten-year period. Long term funding allowed local partners the time to develop and amend local programs and to solve implementation problems. In addition, it focused the attention of the partners on establishing the long-term working relationship that is essential for cooperation. HUD was reluctant to abandon the relationship it established with its partners. HUD monitored performance reports and, when necessary, issued warnings to partners about the need to improve performance.[3] HUD also provided technical assistance (HUD, 2001b).

Although HUD initially emphasized local control in performance monitoring, the need to report progress to Congress created a demand for more standardized reporting. In 1998 HUD implemented an automated performance measurement system called PERMS to monitor EZ/EC communities. In some ways, the design of this system was at odds with the emphasis on local problem solving at the heart of the EZ/EC initiative. According to an evaluation of the system, PERMS is primarily intended to provide HUD with the information it needs to fulfill HUD's legislatively mandated requirement to report on the progress that EZ/EC designees have made in achieving their objectives of community transformation (Elwood, *et al.* 1999).

HUD's desire to manage its partners implementing different programs at multiple sites was complicated by its desire to provide coherent reports to Congress. While the initiative was designed to encourage local problem solving, HUD wanted to summarize progress across sites in Congressional reports. There is tension in these competing purposes. The initiative favors local solutions and creation of local performance indicators to assure that partners follow through with their plans. Standards vary from one site to another, making it difficult to aggregate data and compare across sites. PERMS was designed to encourage uniform reporting across sites so that information can be aggregated and sites can be compared.

The implementation of PERMS created other difficulties for HUD's partners. The system required partners to translate their implementation plans into several interrelated components including baseline conditions, proposed outcomes, milestones, budget information, and output measures. While most participants were able to link budgetary outlays to program activities, many had difficulty translating their implementation plans by the standards PERMS required (Elwood, Hebert, and Amendolia, 1999). PERMS assumed that partners could create a logical connection between current conditions (the baseline), programmatic actions (use of budgetary resources to achieve milestones and generate program output), and improvements in conditions in the zone (outcome measures). Partners were comfortable accounting for financial resources and production of outputs, but had difficulty expressing or creating a coherent logic for their implementation plans.

The difficulty experienced by HUD's partners is related to the management philosophy that is behind performance-based contracting. While public management traditionally emphasizes financial accountability and production of program outputs, performance-based contracting views such concerns as insufficient. Performance-based contracting is part of the ongoing project to rationalize the connection between the means and ends of public policy (Hatry, 1999). The point is to discover the effects of governmental programs, to measure accomplishments, not activities. HUD's expectations were unusually demanding because they required that the linkages between problems, activities, and accomplishments be explicitly drawn. Many local partners were unaccustomed to thinking that way.

A related concern is the usefulness of performance monitoring to local program managers (Newcomer and Scheirer, 2001). The rationalization of means and ends is thought to benefit local program managers (at least those local managers who share the underlying management philosophy). Local managers should find performance indicators to be a useful source of information about program effectiveness to help them identify promising approaches to solving public problems. HUD's desire to create coherent reports for Congress limited the value of PERMS as a local program management tool (Elwood, *et al.*1999). This created two difficulties. First, since the purpose of performance-based contracting in this context is to establish a system for monitoring progress so that partners can work together to identify and solve problems, a disconnection between the performance monitoring system and local programs diminishes its effectiveness. Beyond this, it is difficult to get partners to expend the effort required to support the performance monitoring system if they cannot truly integrate it into program management. This diminishes the value of performance monitoring; rather than being a useful management tool, it becomes a hoop through which partners must jump to satisfy sponsors' demands.

PERMS strives to integrate strategic objectives, implementation benchmarks, and program outcomes. However, to some extent, what is an objective, a benchmark, or an outcome, depends upon your point of view. It is useful to develop implementation benchmarks because initiatives can falter at that stage of the process and a significant amount of time can pass between implementation and outcome. A system that is oriented toward detecting and correcting errors must include implementation benchmarks or problems will not be discovered until outcomes are known and then it is too late to correct mistakes. However, when developing implementation benchmarks it is unclear how to identify important, discrete activities and from whose perspective such activities should be identified. Baltimore's PERMS report provides an illustration. A program called 'Workplace Enhancement' offers training to currently employed zone residents who wish to upgrade their job skills and secure more lucrative employment. Several implementation benchmarks are listed in the report including this entry: 'Signed contract with Johns Hopkins to offer workplace enhancement classes to 35 Zone residents' (PERMS, 2000). While this entry seems a useful implementation benchmark, other possibilities exist because key implementation activities happened after this event (for example, training programs had to be organized and residents had to be recruited to participate). These aspects of program implementation were omitted because the report reflects the standpoint of its authors. Officials and staff at Empower Baltimore Management Corporation, the quasi-public corporation that manages the Empowerment Zone initiative, composed the PERMS report; from their standpoint, the implementation process was largely completed when the contract with Johns Hopkins was signed.

Performance-based contracting can be managed flexibly to allow programs to change as they are being implemented. The 'customized training' program in Baltimore illustrates this. The program provides grants to employers to train zone residents for specific job opportunities. This is viewed as an effective means to overcome a common difficulty in employment training, that participants complete the training only to find there is no demand for their new skills in local labor markets. One unanticipated implementation problem was that zone residents were reluctant to participate in customized training because it was time consuming and many suffered financial hardship during the unpaid training period. Empowerment

Zone officials responded by offering training stipends ($100 per week) and by hiring a staff member to recruit participants. Another unanticipated implementation problem with customized training resulted in the creation of the Workplace Enhancement program. This program was developed when the University of Maryland requested a customized training grant for surgical technicians. Some hospital employees wanted to participate and upgrade their jobs, but the customized training program was geared to recruiting and training the unemployed. Workplace Enhancement allowed zone residents who were already working to improve their skills and job prospects. The University of Maryland Hospital trained 10 participants, some of whom were already employed by the hospital and others were unemployed.

Although performance monitoring can be a flexible management tool, it is no panacea. The value of performance monitoring (when applied to service programs) is limited by the lack of control that partners have over the behavior of target groups. Baltimore's report (PERMS, 2000) frequently states that they plan to "offer" job training or drug treatment or other services to zone residents. This reflects a realistic appraisal of the limitations of their control over the target group. Limited control is a serious problem if performance-based contracts specify performance benchmarks and program outcomes. Either the contract substitutes activities for outcomes (something Hatry, 1999 argues we should not do) or partners are held responsible for outcomes they cannot control. In either case, performance monitoring is problematic.

Performance monitoring is also limited as a management tool because it cannot adequately address questions related to program evaluation.[4] One of the central goals of Baltimore's program is to assist zone residents in finding employment. Baltimore's PERMS report states a "projected output" that 5,000 zone residents will participate in job matches through the state of Maryland's CareerNet job bank (an automated database linking employment opportunities to potential employees). As of August 2000, 4,234 zone residents had participated in this activity. When compared to the benchmark, the program had an 85 percent success rate (4,234/5,000). However, this indicator demonstrates several limitations of performance monitoring. First, the indicator does not identify a baseline participation rate among zone residents, so it is not possible to say whether the number of residents participating is higher, lower, or about what should be expected. Second, even if the performance indicator was modified to include baseline measurement, the question of attributing observed changes to program activities remains. The number of zone residents registering for the state's CareerNet may have increased because of welfare reform, a robust national economy, or even other Empowerment Zone programs.

Even when programs can be linked to outcomes, other limitations of performance monitoring are evident. Baltimore's report (PERMS, 2002) touts the accomplishments of the customized training program: 21 training programs have been completed; 157 residents have been trained and 90 of these have been placed. The report is useful because it informs HUD about the number of residents trained and placed. However, the report does not establish that the program is "worthwhile" from a benefit-cost perspective. One program was budgeted for $280,000 to train as many as 70 zone residents, or $4,000 per trainee. Is this a reasonable expenditure? How does it compare to other training options? How much income gain is expected from the $4,000 investment? Performance indicators alone cannot address these questions.

Another limitation of performance measurement is that it is difficult to sort out the effects of multiple programs. This limitation is especially relevant to the EZ/EC initiative because local programs are supposed to be constructed as a multi-faceted, integrated attack on urban problems. For example, there are several different initiatives in Baltimore's Empowerment Zone that are oriented toward reducing crime. In each case, the crime rate reported in zone neighborhoods is used as a performance indicator. Even if the other possible influences on the crime rate are accounted for (such as changes in unemployment or criminal law), using the same indicator for the same geographic areas to report the accomplishments of multiple programs does not say anything concrete about whether, or which of, these programs work. Hatry's (1999) view that the connection between means and ends in public policy can be rationalized by performance measurement is problematic when applied to the multiple, overlapping programs that exist in Baltimore's Empowerment Zone.

Conclusion

Performance-based contracting has virtues and limitations as a means to manage intergovernmental partnerships. Performance-based contracting encourages cooperation by making the terms and expectations governing partnerships transparent. This eases uncertainty about the consequences of cooperation and encourages coordination (this helps prevent problems that can result from the inadvertent failure of partners to meet the expectations of sponsors). However, partnerships are a special case for performance-based contracting that requires a flexible and dynamic contracting process. Partners form committed relationships in an uncertain environment. The contracting process must allow partners to make constructive adjustments to policy as it is being implemented. While it is necessary and appropriate to specify what partners are expected to do in exchange for financial support, sponsors must allow reconsideration of the plans as circumstances change. Program development is an ongoing process.

The importance of flexibility was evident from the case materials describing Baltimore's Empowerment Zone. Officials in Baltimore frequently revised their agenda, adjusting existing programs and adding new ones, to meet the emerging needs of zone residents and businesses. If too much emphasis is placed upon sticking to initial commitments, performance-based contracting can become an obstacle to constructive policy change. HUD avoided this pitfall by allowing the EZ/EC sites to make adjustments locally so long as these changes were integrated into coherent plans to accomplish the objectives of the initiative. The development of the PERMS system however, was at odds with the EZ/EC initiative, especially the emphasis it placed upon the development of local solutions. Understandably, HUD wanted to provide coherent reports to Congress and the localized character performance monitoring in the EZ/EC program made this difficult. However, the implementation of an automated system that emphasized cross-site aggregation and comparison diminished the value of performance monitoring as a local program management tool.

For performance-based contracting to realize its value sponsors must appreciate its virtues and limitations. Performance indicators provide an overview of program activities and accomplishments. This information should be useful to sponsors and partners alike. However, in partnerships a more detailed understanding of activities is required to assess whether real progress is being made. Here a primary virtue and limitation of performance monitoring is evident. If sponsors receive detailed reports from their partners, the virtue of performance indicators (the summary of progress they provide) is lost in a narrative that focuses on the particular accomplishments or problems of particular programs. Sponsors are likely to be overwhelmed. But, performance indicators alone may mask or distort real performance. Sponsors must use performance indicators judiciously to identify circumstances (potential problems or possible accomplishments) that warrant a more detailed review. When used as a flexible tool to organize and focus the relationship between intergovernmental partners, performance-based contracting has significant promise.

References

Axelrod, R. (1984) *The Evolution of Cooperation*, New York: Basic Books.

Elazar, D. (1972) *American Federalism: A View from the States*, New York: Thomas Crowell Co.

Elwood P. Hebert and J. Amendolia (1999) *Evaluation of the Performance Measurement System for Empowerment Zones and Enterprise Communities*, Cambridge, Massachusetts: Abt Associates Inc.

Hatry, H. (1999) *Performance Measurement: Getting Results*, Washington, D.C.: Urban Institute Press.

Hirschman, A. (1970) *Exit, Voice, and Loyalty*, Cambridge, Massachusetts: Harvard University Press.

HUD (2001a) *Introduction to the rc/ez/ec initiative*, U.S. Department of Housing and Urban Development website. http://www.hud.gov/offices/cpd/ezec/about/ezecinit.cfm.

HUD (2001b) *HUD's role*, U.S. Department of Housing and Urban Development website. http://www.hud.gov/offices/cpd/ezec/about/hudrole.cfm.

HUD (2001c) *RC/EZ/EC laws and regulations*, U.S. Department of Housing and Urban Development website. http://www.hud.gov/offices/cpd/ezec/about/ezecoversight.cfm.

Kettl, D. (1993) *Sharing Power: Public Governance and Private Markets*, Washington, D.C.: The Brookings Institution.

Newcomer, K. and M. Scheirer (2001) *Using Evaluation to Support Performance Management: A Guide for Federal Executive*, The PricewaterhouseCoopers Endowment for The Business of Government.

Oye, K. (1985) 'Explaining Cooperation Under Anarchy: Hypotheses and Strategies', *World Politics* 39 (1): 1-24.

PERMS (2000) *Empowerment Zones/Enterprise Communities Annual Report: Baltimore, Maryland Empowerment Zone (2000)*, U.S. Department of Housing and Urban Development website. http://www5.hud.gov/urban/perms/perms.asp

Radin, B., R. Agranoff, A.O'M. Bowman, C. Buntz, J.S. Ott, B. Romzek and R. Wilson (1996) *New Governance for Rural America: Creating Intergovernmental Partnerships*, Lawrence, Kansas: University of Kansas Press.

Stoker, R. (1991) *Reluctant Partners: Implementing Federal Policy*, Pittsburgh, Pennsylvania: University of Pittsburgh Press.

Stone, C. (1980) 'The Implementation of Social Programs: Two Perspectives', *Journal of Social Issues*, 36 (4): 13-34.

Stone, C.N. (1989) *Regime Politics: Governing Atlanta 1946-1988*, Lawrence, Kansas: University of Kansas Press.

Notes

1 The term 'routine' is used here to distinguish purchase of products such as office supplies from more specialized procurement such as defense systems for which supply-side competition is limited and the government is the sole buyer.

2 The author has been associated with federally sponsored evaluations of Baltimore's Empowerment Zone and has done fieldwork there for several years. This analysis, except when otherwise specifically referenced, is his interpretation of events based upon news reports, document analysis, personal interviews, and participant observation.

3 HUD officials met with representatives from Detroit, Chicago, and Atlanta to express concern about lack of progress (see HUD, 2001c).

4 Newcomer and Scheirer suggest that performance measurement should be integrated into a larger system of program evaluation (2001).

Section Four:
Reflections on Participatory Governance

Chapter 14

Interactive Public Decision-Making in Civil Society

W. Robert Lovan, Michael Murray and Ron Shaffer

Introduction

The contributions to this book comprise a diverse set of narratives whose authors are linked by the shared bond of first-hand experience. In other words, each person has 'been there'. That this should be so is almost axiomatic given the central theme of participation in each chapter. This involvement is very much reflected by our editorial style which has not sought to homogenize the presence of individual agency into 'the third person'. The tensions in planning, conflict mediation and public decision-making watermark these essays. But they have also posed deep challenges for their authors who throughout each engagement have retained their confidence in the paradigm of participatory governance, while at the same time being prepared to reflect critically on how things might be done better and differently. Jim Cavaye is his chapter on the Australian experience has neatly summarized the recent provenance of participatory governance as a progression from the welfare state, through the contractual state to the enabling state. Government more widely is now attempting to build a facilitation and partnership role with communities that adds real value to public policy interventions and community outcomes. In other words, the dynamics of the business of governing are dramatically changing, democracy is being confronted by the need for greater transparency and the relative power position of citizens has now made it important that their voices are heard. This book celebrates a significant progression along a route from which, we would argue, there can be no turning back!

In this concluding chapter it is our intention to revisit the model of participatory governance set out in Chapter One and to consider more fully our conception of civil society as a mediating space between multiple stakeholders across government, the associational sector and business. This analysis is further developed as a set of core principles which we believe are central to the way that interactions should be opened up, relationships should be developed, and policies should be formulated and delivered. Finally, we consider how the practice of participatory governance can be deepened through a more explicit identification with arguments of equity.

Public decision-making in civil society

Civil society is a phrase with multiple meanings which, we argue, can also be regarded as simply that space where the interactions between a state sector, a business sector and a civic or associational sector can and do take place. However, the concept of civil society should not be overloaded and in this context it does not include, for example, everything that takes place within or outside the administrative actions of government. At the same time as suggested by Anheier *et al* (2001) civil society is not about minimizing the state, but rather increasing the responsiveness of political institutions. Participatory governance, as the chapters in this book demonstrate, is central to that project.

The important point is that contemporary, democratic and capitalist societies (both at advanced and transformational stages) are marked by a complexity of relationships and behaviors between this trinity of stakeholder categories on public policy issues. Thus, for example, Kingdon (1995) in his analysis of federal agenda setting in the United States notes that the many participants within any arena can be grouped into a visible cluster of actors who receive substantial media and public attention, and a hidden cluster of researchers, bureaucrats, staffers and administration appointees. Interest groups concerned with agenda setting and alternatives specification, and whose activities may be more or less visible, travel between these two clusters. What makes this insight relevant to our model of civil society is that participatory governance is equally marked by considerable learning and movement by actors who operate on the basis of knowledge of the structures. There is an appreciation that policy selection cannot be reduced to an 'either/or' choice, but is required to engage with multiple alternatives. Conflicts between actors within and across sectors about what can work, what is right and what is efficient draws in the need to operate resolution processes which can connect through to new arrangements for cooperation, coordination and collaboration. Essentially, therefore, responsive public decision-making requires the creation of bridges between top-down and bottom-up perspectives on policy formulation and implementation which, as a longstanding dialectic, has been well rehearsed within the field of planning theory (Murray and Greer, 2000).

The history of planning in the public sector over the past forty years has been marked by two main tendencies, as indicated by Healey *et al* (1982). On the one hand, there has been a tendency to centralism, to depoliticising decision-making and increasing the role and power of technical experts - an essentially managerial tendency with an emphasis on implementation. On the other hand, there have been demands for more participation in decision-making, a call for more accountability on the part of local politicians and officials, and increasing criticism of technical expertise; in effect these are demands for increasing politicization of decision-making. These two tendencies, which are very much at odds, one with another, have been labeled as the top-down and bottom-up approaches to planning, certainly as it relates to public decision-making.

The top-down approach can trace its heritage to the rise of intellectual technology, not least in terms of systems analysis, in the 1960s. Previously planning was theorized in terms of knowledge about the phenomenon, for example, how did cities work, or what were the dominant uses of rural areas. Whatever the benefits, and there were many, of this substantive approach to theory, it tended to flounder when brought to bear as the medium for introducing and managing change in the increasingly complex regional and local socio-economic conditions that marked the early 1960s. The concerns of this substantive approach are well captured by Keeble (1952) in his definition that:

> Town and country planning might be described as the art and science of ordering the use of land and the character and siting of buildings and communication routes so as to secure the maximum practicable degree of economy, convenience and beauty. (p1)

The major criticism of the substantive approach was that it offered little by way of thinking about how planning should be approached, and it was virtually swept away by the introduction of Procedural Planning Theory, whose core message was that what was needed was a theory of planning itself, before its application to or in any substantive area. The theory, born out of systems analysis and related fields such as linear programming, appropriated the term *Procedural* because it reduced the subject area to a number of seemingly unproblematic linked steps which could be listed as goal setting, survey/data gathering, forecasting, strategy preparation, evaluation and implementation, with a cybernetic 'heart' that emphasized constant monitoring of the system to regulate change.

This proceduralist approach appealed enormously to policy makers more generally as a clear cut technical process that laid bare the essence of topics and relegated the 'judgment calls' and 'gut instincts' of politicians to the sidelines. It quickly became a toolkit of virtually universal applicability in the public sector, with its contentless and context less overtones as described by Dror (1963):

> Planning is the process of preparing a set of decisions for actions in the future, directed at achieving goals by preferable means. (Dror, 1963, cited in Faludi, 1973, p330)

The general embrace, afforded across the spectrum of policy sectors, allowed it to assume what is now commonly referred to as paradigmatic status after Kuhn (1970). While the term is now in general coinage its full ramifications in the Kuhnian sense are still, perhaps, not fully appreciated in that Kuhn looked upon a paradigm as an interdisciplinary matrix of theory, shared beliefs and values, and a common repertoire of problem solutions that bind a scientific or technical community together.

However, the clinical, technocratic planning world of the 1960s, when Procedural Planning Theory enjoyed general, if not fully understood acceptance, was short lived and the subsequent history of planning theory is one characterized by a splintering of this once all pervasive approach. The 1970s and 1980s saw attacks from left wing Critical Theory and the New Right, to name but a few, but while these took their toll on the overall structure of Procedural Planning Theory, they tended to have a shallow impact, not least because of inherent faults in their own analysis. Much more effective was the 'slow burn' of Humanist Theory as promoted, for example, by Friedmann (1973) with its emphasis on the societal ramifications of planning and the relationship between experts and communities. The rise of bottom-up planning is based, surprisingly in the context of the longstanding mission of planning as an agent of social reform, on the central importance of communication (written, verbal, street protests, etc.) between planners and those for whom the plan is intended. This communicative or participatory planning has been underpinned by the phenomenal growth of the associational sector. As argued by Scimecca (1995), the starting point of people being limited by social institutions creates the corresponding goal of seeking change in the status quo. The humanist perspective searches for points of effective intervention in the structures that shape human lives.

The whole edifice and ideology of traditional governance has been challenged to accommodate this 'new' societal construct. Thus in terms of land use planning, for example, theorists such as Healey (1989, 1992) and Forester (1989) have been at the forefront of the changed circumstances which are now so prevalent as to deserve the accolade of the 'new planning paradigm'. The spirit of this approach is admirably encapsulated in an early statement by Healey (1989):

> Environmental planning involves the relation of knowledge to action in the management of environmental change ... This would recognize the diversity of legitimate interests in environmental change and the role of discussion and negotiation as the media through which knowledge is translated into action, so ... what is proposed is a form of planning as debate. (Healey, 1989, p7)

The above quotation is the very essence of reasonableness. Its rootedness in humanist theory suggests, for example, that if individuals are given reliable information upon which to make judgments, are provided with the circumstances within which judgments can be made, and are given opportunities to implement their decisions, then the majority would espouse humane values and noble causes (after Scimecca, 1995). However, this position neglects the *real politic* of power relations and, in particular, the nature of the interactions between central government and local communities. Bluntly put, it assumes an equality or at least a structured continuity of relationship between the two that has seldom existed in practice. That admitted, a growing number of chinks in the wall between the bastions of central government and the besieging masses of community interests are now beginning to develop. This reality is reflected in the chapters of this book and more generally by the growing number of recognized public interest advocacy groups and non-governmental organizations and, of course, the growth of innumerable partnership arrangements between central government and local government with the associational and business sectors. This can be regarded as the hallmark of a new participatory governance over the past decade.

However, the process of paradigm change is uneven. While the Procedural Planning Theory approach has come under severe criticism in terms of its indifference to local communities, the asymmetrical impacts of resource allocation, the overblown reputation of experts, and the extremely difficult issue of describing, let alone providing boundaries to the 'fuzziness' of societal systems, it retains a hold on the methodology of planning in the public sector. This is a crucially important point. For all the incisiveness of alternative approaches to planning and their critique of Procedural Planning Theory, they have not developed alternative credible methodologies. This remains the preserve of the centralist step-by-step approach, elevated to the glamour of an allocative model by government.

Nonetheless the integrity of the bottom-up approach remains. Thus, the central issue is methodological in nature, which is a *sine qua non* of paradigms. In other words, those involved have to sign up to a general approach to issues. The fundamental question may be baldly stated as "How can the approaches of the top-down stances of central government and the bottom-up stances of communities be reconciled?" The model of interactive participatory governance outlined in Chapter One is our response to that fundamental question. The contributions to this book examine the institutional and operational opportunities for this style of public decision-making in civil society with the many insights which are offered, capable of being distilled into a set of guiding principles.

Participatory governance principles

Undoubtedly it is much easier to list a suite of principles than it is to build effective action around them and to keep them in place. This reality should not, however, undermine their value, but rather make those with a brief and interest in participatory governance more sensitive and creative about how they are devised and applied (after Haughton, 1999). The case study material explored in this book is located within contexts of diversity and transformation which span political histories, political systems and political actions. The challenge, therefore, is to devise a series of participatory governance principles at a sufficient level of abstraction that they can command some measure of universality. Nevertheless, it is appropriate to remember that key principles are not static entities; indeed they evolve over time as understanding of our political, economic, social and institutional context changes and of course, as this context itself changes. On the basis of the evidence presented in this book, we suggest that participatory governance can be defined against seven interlocking criteria which serve to inform how this engagement process can translate into more effective public decision-making. Each proposition will be briefly discussed with reference to the foregoing chapters.

Participatory governance is integrative: public decision-making concerns itself with making connections vertically and horizontally between multiple stakeholders. Interdependency between top-down and bottom-up perspectives is vital in securing overarching policy contexts and devolving responsibility for implementation. This integration seeks to enhance the capacity for seamless policy-making and smooth management towards agreed outcomes.

Nevertheless, as hinted at in the section above, the hegemony of the top-down can be difficult to moderate and there is an easy option for government to limit the objective of community engagement to merely listening better. Jim Cavaye in his chapter on the Australian experience of community engagement challenges this behavior by contending that interdependency between government and community members allows for better management of dilemmas, more effective coping with risks, policy experimentation and the implementation of tailor-made approaches. Intrinsic to this adjustment are the necessary cultural change processes within agencies to develop the principles and norms that provide a context for more integrative public decision-making.

Participatory governance is strategically driven: public decision-making, in order to be effective, rises above an association with a series of *ad hoc* initiatives. A clear direction, based on an understanding of local issues and supported by a confident but realistic vision of a desired future, is necessary. In conceptual terms the challenge is to maximize the degree of fit between planning and delivery. However, the strategic planning process which is embedded in this principle must be more than a procedural hoop through which participants jump in order to claim the reward of public funding. Strategic planning within the context of participatory governance is inextricably linked to stakeholder empowerment where the benefits go well beyond the planning phase into implementation.

This is precisely the argument made by Norman Reid and Cornelia Flora in their examination of the Empowerment Zone and Enterprise Community (EZ/EC) initiative for 58 rural communities in the United States. Key components of many community-developed strategic plans comprised high levels of community engagement and new leadership behaviors which recognized the program as a community-building process rather than a grant. Their analysis shows that the least was achieved where minimum attention was paid to wide based strategic participation in the formulation of the EZ/EC proposal, to naming the project leadership, and to the perception of the program as a participatory development process involving the creation of partnerships.

This thesis resonates loudly with the conclusions reached by Etienne Nel in his chapter on institutional reform and economic transition in South Africa. Notwithstanding the very significant policy shifts introduced with the dismantling of apartheid, he concludes that the failure to create specific mechanisms for participation and real capacity at local government level, together with serious financial constraints, will impact negatively on the pace and prospects for progressive development.

Participatory governance comprises joint working: public decision-making requires the involvement of multiple stakeholders working together rather than on an individual basis. It is an inclusive activity which embraces the volunteerism found in the associational sector, public officials, elected representatives and business sector participants. Cooperation, coordination and collaboration represent different levels of partnership interaction with the adoption of any mode highly dependent on the history and politics of inter-organizational working, agenda complexity, and a capacity for creative thinking.

Joint working can require formal institutional change as illustrated by Jeanne Meldon, Michael Kenny and Jim Walsh in their chapter on local government, local development and citizen participation in Ireland. From a position of having engaged citizens only in limited passive participation, people are now being given new opportunities to have a stake in collaborative structures and practices concerned with the planning and delivery of programs and projects. The key insight from this chapter is that representative democracy has the potential for considerable enrichment as a result of some degree of institutional *rapprochement* with participatory democracy. Strong leadership by government is a necessary component in helping to facilitate this union.

Participatory governance is multi-dimensional in scope: public decision-making in this sphere embraces a wide range of often inter-related concerns. It does deal with business growth and job creation, but at the same time can extend across to a wide range of social action. Participatory governance can reach out to the most marginalized in local societies but, in so doing, this also requires the involvement of those who may, in relative terms, be more asset rich. It is appropriate, therefore, that this book should include a perspective from faith communities whose mandate is to facilitate a society in which mutual respect, understanding and compassion are hallmarks. The chapter by Randall Pryor provides concrete examples of how faith-based social service programs impact the public sector and remain integral to the call to service inherent in their religious tradition.

Making places work better is also a key feature of participatory governance and thus physical regeneration, along with infrastructural and environmental improvements must frequently parallel economic and social investment decision-making. This multi-dimensionality is well illustrated by Robert Lovan in his chapter on inter-jurisdictional transportation planning in the Washington, DC area. The data confirm that any process to improve transportation must include a strong focus on providing access to jobs, services, shopping and recreation and should, therefore, rise above the constraining threshold of solely highway projects. Transportation planning, in other words, must integrate with land use, environmental and economic planning. The author recognizes that collaboration driven leadership at the top of the policy hierarchy is required, but that this in turn must be rolled out to embrace business, environmental and civic groups in a multi-faceted consensus-building process.

The argument that participatory governance must embrace a multi-dimensional agenda is also underscored by Christopher Rice and Carol Kuhre in their chapter on the work of Rural Action in Appalachian Ohio. This narrative, as with many of the others, throws several principles into sharp focus, but what is worth highlighting here is the connection forged between sustainability and health. Deliberations between citizens, academics and government officials have led to an interpretation of this vocabulary to mean healthy community, healthy economy, healthy environment and healthy individual. The strategic initiatives pursued by Rural Action are located within this understanding of inter-connectedness. At a time when public policy is more generally criticized for its retreat into silos of decision-making, this chapter sends a strong message of support for those who champion the social and political interconnectedness of organizations and agencies which work with interconnected communities, economies and ecosystems.

Participatory governance is reflective: good public decision-making is always willing to learn from experience regarding what works well under different circumstances and what could work better. Monitoring and evaluation can strike a combination of terror and frustration into those using public money, and while accountability and probity of spending are necessary, it is frequently the case that sufficient opportunities for open reflection and shared learning do not occur. Moreover, monitoring and evaluation can be highly participatory by allowing service users, for example, to determine the choice of indicators and not just to supply data. These activities can also be pluralistic by recognizing that there are multiple and, often, competing interests associated with or impacted by decision-making.

The operation of monitoring and evaluation within the context of participatory governance must, however, go beyond financial accountability and concern with programmed outputs. In this vein, Robert Stoker in his analysis of the promise of performance-based contracting in Baltimore's Empowerment Zone argues the importance of establishing a system for reviewing progress so that partners can work together to identify and solve problems. Performance-based contracting allows participants to make constructive adjustments to policy as it is being implemented. Program development is thus an ongoing process and while it is necessary and appropriate to specify what partners are expected to do in exchange for financial support, the point is emphasized that sponsors must allow scope for the reconsideration of plans as circumstances change. In short, the chapter makes the case for managing better the relationships among intergovernmental partners through the flexibility offered by a more reflective approach to monitoring and evaluation.

Participatory governance is asset based: while funding is crucial for the processes and interventions of public decision-making in civil society, the shared assets of stakeholders in terms of knowledge, experience, skills and networks are profoundly more significant for sustainable success. This suggests that participatory governance is more directly concerned with what can be achieved given resource and implementation mechanism realities.

This focused transformational power of participatory governance is well captured by Richard Gardner and Ron Shaffer in their discussion of the National Partnership for Rural Development in the United States. They point out that the National Partnership does very little to alter the resources and market modes of the policy framework, but rather addresses the political context, as well as the institutional structure and behavior in which resources and markets are brought together. The translation of this effort to State Rural Development Councils through diversity of participation has allowed for the vast amount of insight and resources embodied in rural people to be employed in building a common strategic decision-making framework among members.

Participatory governance champions authentic dialogue: authentic dialogue allows citizens holding diverse perspectives to engage with each other in inquiry, sharing and learning. It can allow service organizations and the people with whom they interact to begin to discover principles for a more productive engagement out of which can flow new practices of understanding and respect. Authentic dialogue comprises a number of dimensions which can be summarized[1] as follows:

- it is a process with different interests working together towards a common understanding;
- in dialogue it is necessary to listen to the other sides in order to understand, discover meaning and search for agreement;
- dialogue exposes hidden assumptions and causes reflection on one's own position;
- dialogue has the potential to change a participant's point of view;
- dialogue creates the possibility of reaching better solutions / strategies than any of the original proposals;
- in dialogue participants are asked to temporarily suspend their deepest convictions and to search for strengths in the positions of others;
- dialogue can create a new open-mind set of attitudes: an openness to being wrong and an openness to change;
- dialogue should generate a genuine concern for other persons and thus not seek to offend or alienate;
- dialogue works on the basis that many people have pieces of the desired ways forward and that by talking together they can put them into a workable set of strategies;
- dialogue does not call for conclusions to be reached, but rather is an ongoing process linked with the development of a capacity to think of new strategies and to take sustained actions.

In short, the contributions to this book illustrate time and time again that dialogue has the potential to create new contexts, new overarching values and new future directions. Frank Dukes in his discussion of the alliance forged between the tobacco farm and public health communities articulates very well the power of authentic dialogue. The agreements reached did not arise because the two sides wanted to be nice to each other. Nor did they reflect a relatively common temporary compromise between continuing adversaries. Rather, the key insight from this chapter confirms these agreements as representing a fundamental realignment of interests embedded in an improbably strong alliance, developed from deliberate efforts to foster constructive dialogue and to create productive relationships around mutuality, integrity and respect.

Authentic dialogue has the capacity to mediate division at different spatial scales. Thus the chapter by Michael Murray and John Greer demonstrates how a process of participatory regional planning has been able to flourish in the deeply divided society of Northern Ireland where space is sharply contested. Alongside necessary constitutional agreement and institutional reform (in their own way a product of dialogue), the preparation of an agreed land-use strategy has been facilitated by central government through processes of shared learning based on conversations which have engaged new community coalitions, the business sector and local government alliances. At a more local scale the chapter by David Deshler and Kirby Edmonds discussing conflict management and collaborative problem solving within a protected area of Ghana similarly endorses the efficacy of well managed dialogue for fostering the type of social learning that can lead to conflict resolution. The public examination process in Northern Ireland and the workshop format adopted in Ghana are arenas for learning within which shared knowledge emerges as a more effective option than the imposition of solutions imposed from the top-down. Getting the balance between what works, what is right and what is efficient, as set out in our model of interactive public decision-making in civil society, has been a common achievement.

Conclusion

Participatory governance is now part of the mainstream! Because public sector resources and capacities are now inadequate to the scale of public problems, solutions also require the mobilization of effort from the business and associational sectors. Accordingly, the principles cited in the section above, not only are descriptive of what has been happening in each of our case studies, but also have a normative significance in signposting the way forward for a more responsive polity. In bringing this book to a close, there is, however, one additional matter which we believe deserves to be highlighted as a meaningful contribution to the deepening of participatory governance. The initial part of this chapter focused on the need to mediate the enduring tension between top-down and bottom-up public decision-making perspectives. In our view this can be only be achieved through ensuring that greater attention is given to equity and distributional objectives in all aspects of policy formulation and delivery.

It has long been recognized, for example, that economic growth and efficiency do not operate in isolation from the distributional outcomes, which in turn may have damaging effects on the ability of markets and localities to perform effectively and deliver jobs and wealth. Distributional outcomes from the economic sphere may lead to greater marginalization of particular groups through being laid-off and long term unemployment. This, in turn, may reduce the ability of these groups to gain access to the labor market due to their lack of skills and motivation. The knock-on effects of low income and poverty and the reduction in size of social networks create additional barriers for these groups to climb over as they seek to move towards economic independence. Furthermore, the detachment of individuals and communities from the formal world of work may increase the costs of doing business as the private sector is confronted by recruitment and retention constraints. Considerations of social capital and the wider issue of social inclusion become significant under these circumstances and thus must be incorporated into any understanding of the development process at national, regional and local levels. This is precisely what the contributors to this book have been promoting in their praxis of participatory governance.

Moreover, alongside the distributional outcomes from the development process, there is a clear need within the mediating spaces of civil society to demonstrate that there is no bias, or perception of bias, in the allocation of resources between localities and communities. Citizens suffering from low incomes, poor housing, lack of access to services and detached from the formal world of work should not be further disadvantaged as a result of their location or community affiliation. Indeed as observed by Katz (2001) in his treatise on redefining the American welfare state, 'families and individuals in need of public assistance, Medicaid, workers' compensation, or unemployment insurance do find it much harder to survive in some places than in others' (p355). At a broader scale, participatory governance has the capacity to challenge such inequities by asking those difficult questions relating to the allocation procedures of the center, by ensuring that the nature and scale of policy omission are understood, and by helping to frame the shape of more appropriate policy interventions. It is, perhaps, by these activities that the credentials of participatory governance should ultimately be judged.

References

Anheier, H. Glasius, M. and Kaldor, M. (eds) (2001) *Global civil society*, Oxford: Oxford University Press.

Berman, S. (1996) 'Paper for Group of Boston Chapter of Educators for Social Responsibility', cited in *National Civic Review* (2000), Vol. 89, No. 1, p. 65.

Dror, Y. (1963) 'The planning process: a facet design', in Faludi, A. (ed.) (1973) *A reader in planning theory*, Oxford: Pergamon Press.

Forester, J. (1989) *Planning in the face of power*, Berkeley: University of California Press.

Friedmann, J. (1973) *Retracking America: a theory of transactive planning*, New York: Anchor Press.

Haughton, G. (1999) 'The future of community economic development', in Haughton, G. (ed.) *Community economic development*, London: The Stationery Office, pp. 225-233.

Healey, P. *et al* (1982) *Planning theory in the 1980s*, Oxford: Pergamon Press.

Healey, P. (1989) *Planning for the 1990s*, Working Paper No 7, Newcastle: University of Newcastle.

Healey, P. (1992) 'Planning through debate: the communicative turn in planning theory', *Town Planning Review*, Vol 63, No 2, pp143-162.

Katz, M.B. (2001) *The price of citizenship*, New York: Henry Holt and Company.

Keeble, L. (1952) *Principles and practice of town and country planning*, London: The Estates Gazette.

Kingdon, J.W. (1995) *Agendas, alternatives and public policies*, New York: HarperCollins College Publishers.

Kuhn, T. (1970) *The structure of scientific revolutions*, Chicago: University of Chicago Press.

Murray, M. and Greer, J. (2000) *The Northern Ireland Regional Strategic Framework and its public examination process: towards a new model of participatory planning?*, A Research Monograph, Belfast: Queen's University, Rural Innovation and Research Partnership.

Scimecca, J.A. (1995) *Society and freedom: an introduction to humanistic thinking*, Chicago: Nelson Hall Publishers.

Note

1 This list is adapted from Berman, S. (1996) Paper for Group of the Boston Chapter of Educators for Social Responsibility, cited in *National Civic Review* (2000), Vol. 89, No. 1, p. 65.

Index